PREDICTABILITY

Our Search For Certainty
in an Uncertain World

SIMON KING

Published by

UNDER PRESSURE
— PUBLISHING —

Copyright © 2018 Simon King

This edition copyright © 2018 Simon King

All rights reserved. No part of this publication may be reproduced, distributed, stored or transmitted in any form or by any means, including photocopying, recording, or other electronic or mechanical methods, without the express written permission of the publisher, except in the case of brief quotations embodied in critical reviews and certain other non-commercial uses permitted by copyright law.
For permission requests see **www.predictabilitybook.com**

978-1-9996114-0-8

Published by Under Pressure Publishing, London

Cover design and typesetting: Sarah Smith Design

CONTENTS

5 PART 1 – HOW PREDICTION STARTED
- **7** Chapter 1 – Introducing Prediction
- **23** Chapter 2 – Religious Times

47 PART 2 – ECONOMIC FORECASTING
- **49** Chapter 3 – Trade and Money (The New Religions)
- **63** Chapter 4 – Trade Means Business
- **73** Chapter 5 – The Science of War and Weather
- **89** Chapter 6 – The Age of Enlightenment
- **99** Chapter 7 – Factories, Workers and Commerce
- **115** Chapter 8 – The Study of Work
- **135** Chapter 9 – Homo Economicus
- **155** Chapter 10 – Economics, Politics and America

169 PART 3 – BEYOND ECONOMICS
- **171** Chapter 11 – Predicting Beyond Economics
- **189** Chapter 12 – The Business of the Mind
- **205** Chapter 13 – A Crisis of Prediction
- **223** Chapter 14 – The Undermining of Economics
- **239** Chapter 15 – Predicting Behaviours
- **257** Chapter 16 – At The Forefront of Technology
- **281** Chapter 17 – Measuring Humans
- **299** Chapter 18 – Data Has Its Price

327 PART 4 – PREDICTING THE FUTURE
- **329** Chapter 19 – Visions of the Future
- **349** Chapter 20 – Who Governs in the Digital World
- **369** Chapter 21 – Automated Futures
- **387** Chapter 22 – Where We Are Now
- **405** Chapter 23 – Technology and the Future of Prediction
- **417** Chapter 24 – Staying Unpredictable

- **435** Acknowledgements and Thanks
- **437** Notes and References
- **441** Index

To Charlotte

PART 1

HOW PREDICTION STARTED

To consider how and why humans have sought to predict the future is to consider how society itself has evolved. It is not enough to know what has happened, or what is happening; true power belongs to those that know what will happen. Those laying convincing claim to foresight became leaders, or something very close, in early society.

The need to forecast weather, war and fortune, the vital elements of life, was in the hands of a learned elite with claim to spiritual, mystic power. Later religious learning and divine understanding took over, but the issues remained much the same.

Whilst all that concerned society was harvests, who would make the best leader, and the occasional skirmish with a neighbouring settlement, prediction was confined to only a few areas of life. Humanity came to dominate the world by being curious and by solving the problems it created. Curiosity would lead to geographical exploration. Problem solving would lead to money. Both would make accurate prediction more important and more difficult.

Commerce and trade became the dominant forces in the world; the economy became the main concern of political rulers. The Industrial and Scientific Revolutions reduced the reliance on religion whilst also redefining how prediction was carried out and on what aspects of life. Prediction moved from the mystical and the spiritual to the philosophical and increasingly the technical. People also gained more power and influence over their political rulers – whether those rulers were elected, hereditary, or imperial. Popular power just added to the complexity and the demands on those that predicted.

1

INTRODUCING PREDICTION

The movement of the stars. The patterns in the entrails of slaughtered animals. Comets, weather events, volcanic eruptions, communing with gods and with the dead, reading and interpreting ancient texts. By studying these things, the mystics, priests and learned figures of history would predict and advise. Their predictions would inform everything from when to plant and harvest crops to when to wage war and with whom.

These influential figures, with the ears of rulers and the wealthy, dazzled observers with their command of arcane language and bewildering formulae. They themselves were made wealthy and powerful by their predictions and their ways were closely guarded. In time, the power of these figures gifted with supernatural insights gradually gave way to science, measurement and mathematics. Disciplines those sages themselves had sometimes applied to make their predictions. The complex maths used to track the stars is the same maths used to find statistical patterns in anything whose change can be measured over time. The multiple relationships between observable events, the effects of trade, migration or law on the wealth of a nation, became the terrain of the economist.

For those in charge the ability to see into the future, to be sure that your action was right and would yield rewards, that mystical forces would favour your chosen path, was a vital advantage. Whether waging war, ruling a disparate people, or planning your finances, having better foresight than the opposition was crucial. It was the difference between success and failure; losing and maintaining your position.

Whilst rulers moved from relying on spiritual, mystical or religious predictions to those of economists and other social scientists, the people looked to culture and entertainment. Art, so often looking to historical or contemporary events for inspiration, also started to look forward. Early fantasies of human flight, time travel or other worlds combined with real scientific advances to reflect on what might happen years, centuries into the future. Speculating on what the future might look like, often with a political or ideological critique or warning, became an ever more popular art form. There were visions of a perfect world (Thomas More's *Utopia*) and commentaries on authoritarianism (George Orwell's *Nineteen Eighty-Four*); warnings of the dangers of 'playing god' (Mary Shelly's *Frankenstein*) alongside criticism of class and empire (HG Wells' *The Time Machine* and Jules Verne's *The Mysterious Island*). Amongst the stories and subtexts were visions of how future societies would work, communicate, govern themselves, travel, dress, and wage war.

To add to this a whole industry offering advice to companies and governments on what the future holds for them has developed. Over the last half century a lucrative and competitive market in prediction has emerged, especially since the digital revolution changed how societies both operate and can be observed. Companies seeking to understand disruptive competitors, political shifts, the 'new' consumer and much else have spent millions on outside consultants telling them what is next. Company boards and management, charged with maintaining share prices, nervous and ignorant about events outside their industry and understanding, have adjusted strategy based on plausible speculation because it is better to be seen to be doing anything rather than doing nothing. Governments have shaped policy around short-term, politically-driven priorities whilst at the same time seeking to reassure the public they have long-term plans for the challenges ahead wary of being caught out by unforeseeable events.

How did all this happen? What does it mean? Have any of these assorted attempts at prediction worked? Will they ever work? Will new technologies change prediction?

Humans crave certainty, which they see as synonymous with safety and security. We prefer the familiar. In neuroscience, feelings of uncertainty about what will happen are observed to have a similar effect on the brain as when something goes demonstrably wrong. We fear change, the unknown, because our primitive survival instincts tell us to prepare for the worst. We even express concern about positive change, assuming that it will be overtaken by unwanted consequences (fame brings pressure and expectation; wealth brings responsibilities and difficulties in trusting others). We believe that if something has been around for a long time, largely unchanged, then it must be good; it has proven itself to be the best version possible, so why would we want change?

Yet change is certain and unavoidable. So the only

possible hope is to know what is coming; what the change will look like. But how reliable can any vision of the future be? As the audiences for prediction have become more sophisticated, so the process of prediction has become more complex. But has that complexity made it any more reliable or accurate?

THE ORIGINS OF PREDICTION

Before the written word, before towns and villages or formal structure of law and leadership, humanity and its ancestors were hunter-gatherers. It is impossible to know for sure how our ancient forebears came to learn when and where best to harvest fruit and vegetables or hunt animals. Or if they even understood the effects of rain or sun on the fruition of trees or congregation of animals.

Around 12,000 years ago, humans began to cease wandering plains, savannahs and forests, and started to settle. They built homesteads and farms, developed techniques for growing particular crops. A more dependable source of food was established. At the same time they experienced the impact of previously minor or even non-existent problems. New diseases (of both people, animals and plants – sometimes one and the same) spread more vociferously than ever before. Bodies had to adapt to previously 'unnatural' diets that were now based more around crops and a few, easier to domesticate animal species.

Survival in this world was paradoxically both more reliable and also more precarious. Whilst the attraction of rejecting the nomadic life meant greater control over when and where food would come from, it also increased the all-or-nothing risk. Too much rain or sun, a ferocious storm or natural disaster could destroy a harvest. Two bad harvests in a row or a virulent wave of illness could see a whole village die or be forced to move, potentially bringing conflict with neighbours. Over the centuries the skills required to survive outside of a settlement were lost. The Agricultural Revolution, like every revolution since, brought benefits, problems, and unintended consequences.

On the upside, humanity started to develop skills from house building to baking (it now being possible to grow, harvest, store and process wheat and grains). Tools could now be made with more care and precision, rather than cobbled together quickly on the move. Skills became valued; the highly skilled individual, a significant asset to a settlement.

One of these newly developed skills was the ability to spot patterns; to understand cause and effect. Of fundamental importance to the human dominance over the world was the ability to see not just that heavy rainfall could damage crops, but what conditions might foreshadow heavy rainfall. Forewarned is forearmed. They could plan and pre-empt, they could take precautions, storing more food, harvesting earlier, or move animals to other plains.

The more homesteads, food stores and skilled workers a settlement accumulated, the more others envied them. Defending your hard worked land and produce meant the rise of a warrior class. As a population grew, order was required. People started to ask questions that demanded answers. Decisions on disputes and plans for the future were required and so leaders were sought.

Perhaps most significantly, rules of behaviour and ownership, and of payment for labour and goods, became necessary to avoid or to settle disagreements. To deal effectively with this challenge, records were created and with records came administrators and bureaucracy.

After around 5,000 years of non-nomadic living, writing and counting emerged in the human story. From South America to Asia, there are ancient forms of record keeping the meaning of which is now largely lost. The global growth of fixed populations, authorities and hierarchy, and rules also saw the development of learning, measurement and early science. With them came the rise in importance of an elite, learned few that could record numbers and later laws, stories, names, dates and facts. They would guard their knowledge, pass it on to a select few, and offer their wisdom to those deemed most deserving, useful or profitable.

It is likely that there were select individuals that could better answer questions, impart wisdom, make decisions, and understand (or at least appear to) the world around those few hundred people that made up the village. They rose to influence through ways that cannot now be known, but we can speculate that, in those early times, some element of meritocracy existed. Physical strength may have played some part, especially in times of conflict, but strength alone could not hold a village together and guide it to success. It is just as likely that the clever, the hard working, the capable and those adept at making alliances rose to positions of influence. That this would eventually lead to forms of class structure, closed elites, nepotism and corruption is unfortunate. But those early leaders could, generally, be assumed to have displayed the attributes (not necessarily the attributes we would see in leaders today) that made them best suited within their group to lead.

Over time, those leaders (presumably through trial and error) realised the importance of taking advice. The wisdom of not relying solely on their own instincts, experiences or ideas. They realised that they could be more effective leaders by learning from those around them, and, in some cases, delegating responsibility (and blame). The more effective the leader, the more secure their position and the more the village would flourish. It would grow, repel invaders, conquer other villages, and become a town. The leaders would be more powerful and more secure. However, that success would only come if effectively advised; advised in the ways of war, legislation, agriculture, social order, taxation, and popular opinion. Advised on what we would call luck, fate or fortune. The leaders would make the ultimate decision, of course, but invariably under the influence of various ranks of counsel.

Farmsteads and settlements became villages, and villages became towns and cities. With this expansion war and conflict with neighbours emerged. Conflict would come about usually through the ambition of leaders seeking to increase their power and legacy; for economic reasons to increase wealth or access to resources such as food, metals and so on, especially in conditions of shortage; or to expand the influence of their religion, often under the influence of religious figures.

BELIEF IN A SYSTEM

Difficulty in understanding the world, how and why it works, is a problem as old as human consciousness. Awareness of their place in the world is what marks humans out, along with the curiosity to ask where it all came from. Whilst it is impossible to be certain of any details, increasingly sophisticated beliefs in spirits, the supernatural and deities has been part of the life of humans for scores of millennia. Religions and beliefs were key to answering popular questions of origin (where we and everything else came from), law (what is and is not right) and fortune (why did bad and good things happen without any apparent physical explanation). These all changed with the introduction of static populations. The elements that were vital to existence fundamentally changed and with them the ideas that made them work.

Being in one location made places of worship and holy structures more important. The sky, the mysterious canopy that appeared to bestow so much and change so strangely, was observed always from the same point. The seasons were roughly the same and could be measured. Religion usually centred around some combination of animals and the dead. Astrological bodies became the main explanation for and influence over, some of the most important aspects of life. As such astrology was vital to almost everything and the pre-eminent consideration in decision-making and planning.

The history of the rise and fall of various gods and beliefs is long and complicated, and in many cases, lost to time. Archaeologists have pieced some belief systems together, but mystery still surrounds the purpose of structures, icons and tablets uncovered around the world and assumed to have a religious purpose. One constant source of faith, and one with evidence of application for millennia all over the world is astrology.

Religion in its earliest forms is thought to date back over 300,000 years. Like much of prehistory (history before written records) this is, to a degree, guesswork. Evidence lies in what appears to be ritual burial (rather than simply practical ways of disposing of the dead) and artefacts with no obvious use as tools. Icons like the Hohlenstein-Stadel lion-man found in Germany, or the bones of the pre-historic dead accompanied by beads in Russia, or with animal teeth and shells in Italy. Why would these objects exist? Why would such effort be made, were it not to effect otherworldly powers? We will never know for sure. Controversy hangs over many of the assumptions made and we may never know with any certainty when humanity started making representations to deities, how or why.

Those unknowable things aside, evidence does suggest that religion (as we understand it in broad terms) dating from around 50,000BCE to 5000BCE appears to have centred around the most important and powerful elements of daily experience, most notably animals. It is reasonable to assume that in groups for whom hunting was vital, they would place either the creatures they hunted and depended upon, or the other animals they saw as successful hunters themselves, at the centre of their belief systems. That they would seek to find ways to emulate their success; to think in the way they do or to be imbued by their spirit.

The other great power in the lives of early humanity, especially when hunter-gatherer moved to agriculture, was the sky. Home to the sun and moon, the source of rain; vast and mysterious. Worshipping the sun is one of the best-known and most enduring early beliefs. Whether people understood its life-giving power or not, they were in awe of it. Groups from ancient Egypt and China to the Mayan people are known to have, in some way, worshipped the sun. Again, the precise details of their worship are hard to read, but prayer, sacrifice and ritual are all known to have been offered to the sun.

Around 3000BCE to 2000BCE, the early Mesopotamian and Babylonian civilisations studied the night skies; the moon and the complex set of enigmatic dots that replace the sun. They mapped stars and constellations. The Sumerian people of Mesopotamia developed mathematics that they applied to, amongst many other achievements, this study. Their numerical and algebraic systems allowed them to measure astral and lunar motion. A key to the mysteries of the world had surely been turned.

It is important to consider that during these formative times, science and belief were intimately connected and not generally contradictory. Astrology and astronomy were essentially the same thing. Without an understanding of anything beyond their limited, small-scale geography, it would be reasonable for ancient civilisations to assume that the stars were divine communications to humanity that needed translating. Plotting the accurate movement of stars and assigning meaning to those positions seemed perfectly reasonable.

The most advanced ancient civilisations all developed forms of mathematics and invariably applied it to understanding the mysteries of the sky above. The Babylonians developed a system of mathematics not dissimilar to the Sumerians, but based on the number 60 (60 seconds in a minute etc.) and applied its factors (1, 2, 3, 4, 6, 10, 12, 15, 20, 30 and 60) to many aspects of life's statistical essentials. This included their study of the stars and they divided celestial movement into 12 segments, assigning animals to each segment.

With little or no communication between populations, many concepts and discoveries would remain isolated geographically for centuries. Whilst the peoples of western Asia, southern Europe and North Africa innovated, populations in east Asia and south America went their own way, although in many cases they would draw remarkably similar conclusions. As people travelled, explored, and conquered ever further lands, learning and beliefs intermingled, collaborated, evolved, or in some cases were just imposed upon new populations. Religious beliefs became powerful political tools used to identify a population, generate distrust for certain groups, and impose authority.

Ancient Greece around this time was a land filled with disparate cults and superstitions aimed at explaining the world and its workings. It was chaotic and far from the cradle of advanced thought that it would later be known as. Everything from life, death and injury to harvests, conflict and knowledge was a matter of mysterious forces influenced by varied but distinct practices. It is thought that there were over 2,000 cults practised in ancient Athens alone. Like many superstitions throughout history, some hit accidentally upon practical truths, like washing hands to prevent illness. Some have, strangely, lived on, like cats walking across one's path or a fear of the number 13. All of these rituals and beliefs were suggested, deduced and argued over by dream interpreters, lucky charm vendors and fortune-tellers.

In to this Grecian world of confusion came the 12-segement zodiac (as they would call it) and the study of the stars. By the 6th century BCE it was married to 12 Greek gods, each ruling over an aspect of the world key to the survival and success of the nation – weather, seas, fortune, love, war and so on. As Greece entered its classical period, it started to make the great artistic and scientific strides for which it is famed. It advanced science and mathematics with experimentation, mathematical proofs, calculus and geometry. The gods and the stars, however, remained central.

The Antikythera mechanism is seen by many as the first computer in history. Invented in Greece somewhere around 50-100BCE, its intricate mechanism of cogs and gears was designed to track and predict the movement of the stars, moon and sun. Such incredible work, both of intellect and craft, was only worth doing for something of vital importance to the most powerful and wealthy Greek rulers. The very life of the state depended on work such as this.

The seers and augurs, the priests of ancient Greece and Rome claimed to be able to interpret the intentions of the gods via nature. The flight of a bird could determine the appointment of a civic elder; the slaughter of a goat the start of a war. Likewise the position of the stars when a person was born, or when an expedition was embarked upon, would indicate the gods' approval and therefore likely success. To make things more complicated, gods were not always all-powerful. Others could be enlisted to help a person or enterprise, even human will or guile could usurp divine actions. So the predictive nature of these signs was vital, but not always decisive.

Any question that could be answered in the positive or negative would often be asked of lower ranked priests skilled in the predictive arts. Oracles, the higher rank of priest, communicated directly with the gods and could answer broader questions. The position and standing of the oracle rested on their knowledge of how to read signs as well as their learning from history and their predecessors. This knowledge was guarded jealously. Much of a seer's success would lie in the wording of a question, or more, their ability to suggest what question should be asked, and delivering a carefully worded, often cryptic or ambiguous prediction.

The successful predictor would make sure their successes

were celebrated and their failures explained away as the result of poor questioning rather than poor foresight. Seers and oracles were important, influential advisors to the generals, politicians and merchants that ran cities and states. The people knew that the success or failure of a person or undertaking would be down to a combination of wisdom and chance, and by extension the whim or favour of the gods. The rulers needed access to those whims. A poor ruler or leader would not have been favoured by the gods, whether through bad fortune or because the gods had not bestowed upon them foresight, wisdom or bravery.

The augurs of Rome held similar positions to the seers and oracles of Greece in the determination of state and personal decision-making. They were educated, from the upper echelons of Roman society (although this would change around 300BCE when commoners were permitted to both become and consult augurs), and would study the rules of auspices. Importantly their work was to guide the decisions of rulers. Those in charge of important matters would have a basic understanding of the auspices as they simply judged the wisdom or correctness of a decision, as the gods saw it, rather than directly delivering it. An auspice could be sought by an augur in order to inform a decision, or it could be an event, such as a storm, which would be a sign the gods wished to communicate their judgement.

Most famous of the Greek oracles was Pythia, the Oracle at the Temple of Apollo in Delphi. Seen as the most influential of oracles, the priestess at the temple was given foresight by Apollo, the son of Zeus and the god of prediction. The Oracle would make predictions on the seventh day of the month after bathing in holy water and subject to a range of rites and ceremonies with attendants. A goat would be sacrificed and its entrails examined for favourable signs. Assuming a positive outcome, the Oracle would receive questions from the citizens, with priority given to those representing the elite, and answered from visions she experienced. These visions, it has been speculated on in modern times, may have been due to fumes or poisons present from geological fissures or from the burning of plants that had a role in oracular ceremonies.

Some of the Oracle's predictions, although possibly dating back to before 1000BCE, were recorded from around 600BCE onwards. These predictions were very open to interpretation and rarely a straight answer to a straight question. In 560BCE, Croesus, King of Lydia, a territory east of modern-day Greece, asked if he should attack Persia. The response was that if he did, a great empire would fall. He asked if his rule would last. The response (according to the historian Herodotus), was "Whenever a mule shall become sovereign king of the Medians, then, Lydian Delicate-Foot, flee by the stone-strewn Hermus, flee, and think not to stand fast, nor shame to be chicken-hearted". It could not be clearer. Croesus did not think it possible a mule could ever rule his kingdom, and that the great empire to fall would be the Persian's. Of course, he was defeated in battle (perhaps because his kingdom was led by a mule-like intellect) and his empire was the one to fall. The predictions were correct, but the interpretations were wrong. If Croesus had been victorious, the prediction would have been the same, his interpretation would have been correct, and the Oracle lauded by him.

Apollo was also responsible for the most famous of predictors in mythology, Cassandra. Having the misfortune to have a god fall in love with her, Cassandra spurned Apollo who attempted to seduce her by bestowing the gift of prophecy on her. In revenge for her refusal he added to the gift the curse that whilst her prophecies would come true, no one would believe her. She went on to foretell the fall of Troy and many other tragedies, all of which she warned the world of, but none of which she could do anything about. In time her name would be associated with anyone that offered a warning of future disaster but was ignored.

The future was mysterious, and so it took mysterious ways to predict what it might look like. There simply was no other way. It took god-given talents, intellect and understanding to foresee what only the gods knew and could control.

2

RELIGIOUS TIMES

"You are wearied with your many counsels; Let now the astrologers, Those who prophesy by the stars, Those who predict by the new moons, Stand up and save you from what will come upon you." -

Isaiah 47:13, New American Standard Bible

Since the immergence of language (possibly even before) there has been individuals who claimed to be able to commune with deities and spirits. To converse directly with them, or to be able to derive their whims, approvals and desires through some means or other. From shamans to high priests, an elite, chosen few with the divine gifts of interpreting, predicting and establishing heavenly laws through trance-like states, arcane rituals and ancient texts.

We know of oracles and augers, but according to historians including Cicero, similar practices predated ancient Rome and Greece. Sometimes those learned seers accidentally hit upon conclusions that what would later be realised to have a scientific explanation. Such as the Ancient Egyptian belief that the tidal level of the Nile was a divine sign indicating the productivity of the coming harvest (in fact water levels in this arid part of the world were vital to agriculture).

Sometimes oracles were just plain lucky. It might be that their guesswork, through pure chance, just happened to be right more often than it was wrong. Or perhaps their audience took their prediction, reacted to it in a particular way that also, by chance, just happened to work out to their advantage. Sometimes the oracle was just more astute than most others, and could see what they could not through a logical process of observation and deduction (as a scientist would). Sometimes their prediction might be so vague and carefully worded that it could easily be interpreted as correct after the event regardless of the original intention. It was these skills that really made for a successful oracle. A combination of luck, judgment and subtle language.

With the development of writing and records, predictions and divine portents became a longer-term matter. No longer was it a question of observing a comet or performing a ritual in order to answer a specific question. Signs and forecasts would be chronicled for future reference. As such, prediction became a matter not just of reading contemporary events and signs, and of learning the methodology of foresight, but of interpreting past events and even past predictions. Forecasters would have to place events in a wider, historical context. Some predictions might not come true for decades or even centuries. Reading and learning became a vital part of the prediction business and because of the limited access to education this led to a closed shop of predictors. Access to these scholars of the divine and to their foresight became much sought after, and, as a result, expensive.

Across the ancient world and beyond, spiritual belief systems were being codified and documented. For centuries charismatic leaders and teachers married early scientific understanding and forms of medicine with confident claims of special access to spirits and deities. They developed origin stories often based on tribal myths and oral histories. Rituals and laws were set down by the few capable of doing so. They became leaders and the keepers of traditions and histories that defined a people in a fundamental, emotional way. They offered guidance, consolation, leadership and advice to rulers and to commoners.

Writing replaced the oral tradition of passing on stories and practices. By physically recording the myths and traditions of a group or tribe, they spread further, lasted longer, and changed less than previously. Societies now combined histories (factual and hearsay), tribal rites, moral codes, folklore, remembered and recounted incidents and superstitions. These records took on symbolic meanings of their own, becoming a cornerstone of how a society saw itself and perpetuated its identity. The written records ensured these traditions, cultural identities, beliefs and practices – the customs that would define a population – were passed on through generations, unchanged and sacred. They could be reproduced (by those trained and trusted to do so) and passed to subsequent generations or those exploring new parts of the world.

From these written amalgamations of stories, practices and ideas immerged what would become the foundations and frameworks of institutional religion. Religion as it would be defined for centuries by the likes of Judaism and Buddhism, or by Zoroastrianism – thought to be oldest religion still being practiced today and by most standards the antecedent of many more modern beliefs.

Zoroastrianism combined elements of astrology and monotheism, along with social mores and traditions, and was concentrated in what would become the heart of the Islamic world. It had a central god of creation, defined concepts of good and bad, ideas of free will, the soul, and of life and death. It also established many rules and observances that would figure as tenets in other religions from baptism to concepts of modest attire and behaviour to the importance of good deeds.

Over the course of millennia, the world would come to be dominated by just a handful of religions. The three most influential of these being Judaism, Christianity, and Islam, all of which share common roots, practices, laws and stories. Whilst other beliefs grew and waned around the world, these three led not just personal and communal beliefs, but globally shared ideas of morality and order.

These religions helped establish trust between peoples all over world, whilst also playing a key role in the conquering or assimilating of populations. They offered simple answers to questions commonly asked by the general populace. The oldest of the three, Judaism, would lay the foundations of the other two in its origin story which set out definitive views of how the world and everything in it came to be, the divine gift of man's dominion in the world, his place in relation to God, and the nature of good and evil; the fundamentals of life. They would be known as the Abrahamic religions.

Exception to these beliefs came principally (in terms of numbers) in the form of Buddhism, Taoism and Confucianism – all practical philosophies for life - and Hinduism, which combines a personal and spiritual philosophy with allegorical myths and legends. Despite billions of adherents, the influence of these beliefs stayed, for the most part, geographically confined to Asia. Although important to their populations and serving many of the same purposes as the Abrahamic religions, they lacked some vital element to their wider proliferation.

Perhaps they were not adaptable enough - too unwilling to change to embrace other practises and ideas. Maybe they were too complex for a population already steeped in its own beliefs to simply adopt. It may be that the right people did not practise them at the right time to spread them around the world. Whilst these are all feasible to some extent, the main barrier was that these religions make belief personal; they focus on the individual or the communal. They do not equate moral good with converting non-believers to their cause; they lack evangelism.

The dominant Abrahamic religions have survived thousands of years and have covered vast expanses of the world. They account for the majority of religious faith and they owe much of their success to the development of writing. Trade pre-dated writing, as did religious and spiritual belief. The Abrahamic religions, however, needed the written word to spread their influence and hold on to their adherents. It is why the oldest of these beliefs that is still widely practised, Judaism, is only just over 5,000 years old. Although Judaism in particular owes its survival to more than just writing. It is a religion that survived where other similar religions (by age, by size of following) died out. Partly because it formed the bases of the most dominant, geographically widespread and evangelical religions (Christianity and Islam). Partly because of the suffering its adherents endured. Adherents who, in order to survive repeated attempts to exterminate and assimilate them, held on to their faith and fought fearlessly to preserve it; to preserve their identity not just as a faith, but as a race, as a people. In many ways Judaism replaced the evangelism of other faiths with nothing less than an incredible will to survive.

In 1992, British anthropologist Robin Dunbar, based on observations of primates, postulated that humans could only maintain meaningful, reliable social connections with a maximum of around 150 others. When human settlements grew in size beyond that, challenges to the maintenance of order and cohesion appeared. Language and shared observances, interdependence for food or child rearing would all play a part in stabilising increasingly large groups. Religion, however, not only provided coherent means to explain the world, it created a sense of belonging and trust. It provided social order and stability. It enforced a rule of law few would question by combining social, political and spiritual authority within a ruling elite. In short, it created a culture. Those in charge were often placed there directly by god. Shared religious belief meant peace between groups with little in common, particularly as tribal groups gave way to nation states.

Religion has been spread invariably by wars of conquest, with Christianity perhaps known to be the most successful in this via the Roman Empire. Emperor Constantine converted to Christianity in 312CE; in that he was simply following much of his population, but he set in motion the Christianisation of the whole Empire by his successors throughout the next century. Pagan temples and practices were outlawed at various points up to the 4th century, usually accompanied by driven, often violent zealots, monks, priests and preachers.

As Christianity swept aside paganism across most of Europe it transformed its rites, festivals and observances, adapting pagan elements and moulding them into Christian ones. By the 9th century most of Europe, including Scandinavia and the Balkan region were largely or wholly Christian.

Islam would immerge in Mecca around this time. Judaism and Islam were, like Christianity, born of the same origins in the same geographical part of the world. These two diverged from Christianity mainly around the coming of a messiah; a figure that would ultimately save the world, balance out human sin, and save humanity from evil or lead the good and faithful to a (literal or figurative) promised land.

Christianity believed the Jesus Christ was this messiah (a messiah prophesied in ancient holy texts), whilst the other two believed he was no more than a prophet, albeit a significant one. Christianity, basically a sect of Judaism, placed great emphasis on spreading the word of Jesus Christ as the messiah, making the act of fundamental importance to the faith. Judaism had no such compulsion to evangelism; to spreading its message. The consequences were that Judaism retained a sense of insularity, concerned more with its established followers and the notional land of Israel – meaning that whilst important and geographically widespread, the faith has comparatively few adherents and little modern cultural influence.

The three central texts of these religions, the Qur'an, Torah and Bible, share many stories and characters, particularly when it comes to prophets (although names sometimes change between the texts). Mohammed, to Muslims, is unique amongst other prophets (including Jesus Christ) in being the most holy and the last prophet. Judaism acknowledges 48 prophets (and seven prophetesses), with the last being Malachi (although even his name is disputed amongst scholars) and with whom divine prophecy died.

In religion the prophet is seen as a holy messenger, divinely chosen to impart the word of God either verbally or in writing. As such, what we think of as prophecies only form part of their legacy. These words would take the form of laws and judgements, sometimes advice or reassurance, and only sometimes visions of what would come. It might be how the end of the world would manifest itself, or that the people of Israel would be led through the wilderness, or that a messiah would come, or that holy texts would be corrupted and should not be used.

In this respect such holy predictions were without time constraint (we are still waiting for them to happen), would be proved right only by an event also described by another holy text (potentially written long after the event and its prediction had taken place), or would be so vague or allegorical that interpretations could be argued over for years to come.

Religious prediction has never really had much of the practical, day-to-day function of other forms of prediction for the most part. Religious texts offered all types of guidance and instruction on how to lead a life that would be rewarded in spiritual ways, but their views of the future were less clear. That said, there were stories that proved God's ability to bestow either the gift of prophecy or, more likely, the foresight of a particular event. Famously, Noah was told by God of the coming of a great (if physically impossible) flood.

Around 620BCE Jeremiah, a figure in the Torah, Bible and Qur'an, 'received' the word of God and was told he should dedicate his life to prophetic ministry. The Israelites had largely renounced God and had started worshipping the 'false god' Baal. For this betrayal God would no longer protect Judah. Famine would befall the land, followed by subjugation by a foreign power, and then the exile of its people. Jeremiah preached this vision of doom, but other priests and prophets were angered by his predictions. His words were demoralising and spread panic. Worse, he was blaming this coming catastrophe on the self-serving and corrupt priests. Jeremiah was shunned; beaten and humiliated but retained his faith in his prophecy. The authorities attempted to execute him, only to fail. Inevitably Judah and its capital Jerusalem fell to the Babylonian empire.

Whist history is invariably written by the victors, and frequently not contemporaneously, it is safe to say that the land of Judah did indeed fall under the expansionist Babylonian empire. It is very likely that some committed believers were distressed by the worshipping of different gods and swore that God would punish them (these were very religious and religiously contentious times). The history of believers warning of the consequences of abandoning one god or another is almost as long as the history of religious belief. It is entirely likely some monotheistic believers sought to warn the public of a coming disaster engendered by picking the wrong god and as such enraged the ruling political and religious (then, much the same thing) elite who had a stake in the status quo.

Groups came to define themselves by their beliefs, practices, histories and origin stories. Some deity would usually have created the world, and those in it, and set out some rules which, if observed, would result in good fortune. Breech those observances, in whatever way, and suffer, in this life or in what is to come. These divine stories and rules were, inevitably, written and recorded by and accessible to only a learned elite of religious scholars, elders and teachers. These people were assigned various titles by various traditions, but for the sake of brevity we will refer to them generically as priests. Whilst the texts that set out these stories and laws themselves were often said to be the verbatim word of God, only humans could physically record them. Of course, in some cases (for example, the Ten Commandments), God did physically write (according to some interpretations), but humans copied the text and the original was lost or destroyed, usually in some ultimately meaningful way.

Humans also interpreted the word of God. They judged according to the word of God. They advised according to the word of God. At least what they believed was the word of God. Whilst most religious texts predominantly focused on origin, morality and rites, many also included what would be seen sooner or later as predictions. They foretold of the coming of prophets (those that speak directly the words of a deity) or of cataclysmic events. Many were cautionary – fail to act in a certain way and suffer future destruction or injury. Whether they came from observed history, tradition or myth, inexplicable but clearly significant signs were recorded as divine acts to be interpreted by those to come later.

Some Biblical and Torah predictions (whether they were written as such or not) related to the rise and fall of nations, empires and ruling families. Other predictions would cover the coming of a messiah and the end of the world. Significant events but invariably recorded in less than specific language. As a result all of these predictions have been debated, interpreted a dozen different ways, and put into a variety of contexts that could never possibly have been expected at the time of writing. For thousands of years, and to this day these predictions (some of which were not even written as predictions) have been used to explain global events as the will of God, often acts of vengeance or warning for some perceived human failing. They have been looked to as proof that those recording the word of God were foretelling everything from the establishment of the modern state of Israel to the 9/11 attacks to the clandestine intentions of the UN.

In the same way as the augurs, shamans and those that went before them, the priests and the scholars, the scribes and interpreters of these religious texts were also their guardians. Keepers of traditions stretching generations and protectors of knowledge that few were permitted access to. They would pass down their literacy and learning to chosen successors that they had appointed. Much of their work would be closely guarded secrets, learned over years with details only they understood. They would use this unique knowledge to interpret the will of gods, to lead and to advise.

They would also use their knowledge to secure their position in society, to ensure they remained vital to the rulers and the population that depended on them. They would study the sacred texts for guidance and to uncover patterns and extrapolate from them future events or determine a god's favour. They would also study other works of history, astrology or philosophy, combine them with religious texts and draw yet further conclusions. They would imply consequences and explain the apparently inexplicable work of the gods. Due to their influence, knowledge and stature they would guide and by extension dictate to the masses. Their role within society, their influence at all levels, their power over God made them crucial, even dangerous political figures.

Understanding religion, the language and its ever-growing complexities of interpretation and history required learning. For centuries that meant the priesthood was largely the preserve of wealthy and important families. The earliest educational institutions only taught theological studies, which in some cases expanded into areas of early science and law, alongside languages. This allowed the ruling elite to maintain control over an increasingly important aspect of life. Something that exerted great power over the people, and was integrated wholly into political structures.

Some religious leaders would become political leaders themselves, but more often they would maintain positions as influential advisors to the ruling elite. By parts offering them guidance, offering entreaties to God on their behalf, and endorsing their decisions as approved by God (or otherwise). Priests were heirs to much powerful, sacred knowledge. They would have insight into mysterious events, they would offer access to and even influence over the gods. They would know what was right and wrong, wise and foolhardy. They would understand the nature and intentions of an all-knowing, all-seeing deity. Their teachings and pronouncements were accepted as unquestionable, profound and authoritative, above the tawdry, earthly concerns of politics and immune to the vagaries of popular opinion.

The complexity of the priests' work increased as they measured, calculated, deconstructed, examined and interpreted more and more manuscripts. The central religious texts or holy books would remain, to varying degrees, immutable and preeminent. But as time went on historical chronicles of all types, combined with current events and ideas, effected how certain elements would be seen and understood. For some faiths, other, near contemporaneous texts would be seen as, if not quite equal to their holy book, a very close relation to the divine word of their god. It was for the priests to offer commentary and context, the better to decipher prophecies and divine intentions.

Whilst it was the big issues of messiahs and Armageddon that occupied some theologians, rulers and commoners alike yearned for something more everyday. Religious rulers would be looked to for reassurance as much as prediction, but frequently they were one and the same. Would God look favourably on a marriage or commercial transaction or treaty? Or would it come to haunt those undertaking it? The learning that came with religious authority also brought expectation. If someone were trusted to dictate on matters of morality, if they had the ear of God, surely they would know if a risky enterprise would be worthwhile. Priests may never have sought this role, but they got it, and whilst some may have made their excuses and explained that God did not operate like that, many embraced it.

Arguably the most famous predictor who certainly (rather than possibly) lived, Nostradamus, was a learned and scholarly man. A physician and writer, a man of early science, he combined interpretations of Biblical and other religious prophecy, particularly Armageddon prophecy, with astrology. Even other 16th century astrologers questioned his work, relating, as it did, past astronomical events to earthly events and concluding that the next time such planetary alignments occurred, so would the earthly events. Like forecasters before and since, his gnomic, vague, conundrum-like prophesies have seen them repeatedly interpreted, retrospectively fitted to events to 'prove' their veracity.

History, science, astrology, mathematics, philosophy; they all played a role; they all had an application in the analysis of religious texts. Each offered a new and more convincing result. And with this came ambiguity and vociferous differences in interpretation. The lessons or forecasts, regardless of what may have been intended when written, were defined according to prevailing cultural attitudes and new discoveries. They could be bent to accommodate the desire of rulers. New interpretations would be passed on to future generations as irrefutable, despite their deviation from the original. Old ideas mingled with contemporary ones, unreliable memories of people and events were incorporated, certain lessons and warnings emphasised.

This convolution happened particularly when texts were translated into new languages. Changes were even made in order to make a particular belief system more attractive, to inspire new adherents, or to fit in with new groups in order to expedite assimilation. This complexity, and the death of the languages often used in ancient writings, made the priests' positions ever more singular and necessary.

Whether the guardian of an ancient spiritualism, astrological insight, or a monotheistic religion, being a priest, with all its attendant studying, ministering, answering questions and philosophising, was a full-time job. Only they could carry out the processes, the ceremonies, and the interpretations. Their position within society became vital both culturally and politically. They recorded histories, endorsed ventures, developed nascent sciences, made judgements on right and wrong. Their advice was treasured and incontrovertible. They offered guidance and were the conduit to the deities that would bring good fortune on the people or their ruler (often seen as the same thing). Should harvests fail or battles be lost, someone, somehow, had failed to do their duty by the gods, but it was probably not the priests. If their readings of texts and signs had indicated a positive outcome for an endeavour, then either the outcome needed reinterpreting, or the priests' instructions had not been carried out properly. The prayers, sacrifices and entreaties of those that stood to gain had not been sufficient or sufficiently heartfelt.

The priests' arcane, carefully protected knowledge and dedication secured their position in the political order. They were one of the few constants in most governmental and ruling structures. Their blessing, their predictions, their path to the deity was essential to good rule. A monarch or ruler would have their own, dedicated priest, usually part of coterie of advisers. A hierarchy of priests developed, with those closest to the rulers at the top, which became intertwined with the operation of the state socially and politically, locally and nationally.

With their learning, impact on the popular imagination, and understanding of what was right and wrong, divinely favoured or opposed, priests became vital to effective rule over a nation. As a ruler grew in power, so did those close to them. The priests that advised rulers in turn themselves wielded power. With power comes the desire to protect one's position so a cycle immerged of reserving their roles only for those ordained by God, making religion ever more mysterious and closed to the common person and so their insight ever more special.

Scholarship and dedication to religious understanding were the hallmarks of the priesthood of any religion. They dedicated their lives to understanding, revealing divine truths, and to predicting. No matter how earnest a priest was, though, it would be very hard not to bear in mind the whims and desires of their ruler when advising or instructing. A ruler, chosen by the priest's infallible God, who also called the priest to their vocation. A ruler who almost certainly held the power of life and death over the priest, as well as reward, patronage and prestige. Priests would have to balance what they truly believed their studies and their divine interpretations showed with what their powerful ruler would want to hear.

What did the entrails, stars or sacred texts say about who would win a war, or more likely what favourable or unfavourable conditions would influence a battle? Would the ruler really want to know what the gods said or whether their past actions would inspire divine condemnation? Would the ruler follow the priest's advice either way? If the stars or the ancient texts said the harvest would fail, that the people would revolt, would it be best to say nothing and make excuses later? Say the wrong thing at the wrong time, even if the priest believed it to be true, and they could incur the wrath of an unjust ruler. Would it be worth it?

If the ruler ignored advice, and was removed from power (in some cases permanently) what would that mean for the priestly adviser? Whose side should the priest be on – God or man? Sometimes they would stay on God's side, and sometimes they would pay an earthly price. Make the right call, and survive a little longer; see your position rise as you prove your priestly talents for divine intervention and prediction. Be wrong, and hope your ruler is removed from the picture; you may be able to align yourself with the new leaders. The ability to present predictions diplomatically, vaguely, or with enough room to make excuses later on would undoubtedly be a key skill to both the spiritual and secular advisor to power.

The power and influence of spiritual leaders continued for centuries, only starting to wane in the late 20th century, and even then only in some parts of the world. Their authority did take a serious knock in the 15th century. One of the key advantages the priesthood and the ruling elite maintained was the ability to read and write. Such learning was considered hugely powerful. With the invention of the Guttenberg press, which in turn meant the ability to mass-publish written texts, particularly holy books, learning would slowly become more widespread. People would read the Bible for themselves, no longer reliant on priests. The people would still take the word of the holy men, but increasingly they had the means to challenge their version of God's words. The public started to come up with their own interpretations.

As more people learned to read, so ideas spread more rapidly, and stuck more vehemently. No longer did you need to hear words spoken by an individual in order to understand. People began to think for themselves. And in thinking for themselves, they became more unpredictable. Society became more complex and harder to rule.

THE POWER OF PREDICTION

Prediction can be made easier if you can influence that which you are predicting. Whilst the priest could not necessarily influence the weather or a war (apart from, occasionally, through diplomacy or espionage – especially as religions became trans-national), they could sometimes influence how the people viewed a ruler's decisions or the righteousness of their cause. Priests wielded social influence through a network of clergy and preachers who would pass religious teachings, sacraments and laws to the people. As such senior priests were both advisors and influencers, able to effect large elements of the population as well as the rulers. That meant their position was politically sensitive as well; be seen as too weak or too strong, and a ruler may tire of their presence.

The relationship between God, priest and ruler has a long and complex history that still operates today. Just as priests were chosen through a divine calling and appointed by God, so it came to be believed that rulers were as well. Monarchs, through birth rite, were chosen by God, divinely favoured and gifted with great responsibility and the tools to rule successfully. As such their rule was unquestionable as their decision was the decision of God. Only God would judge them. They did not need the consent or the authority of the people to rule. They were above that.

Ancient Egypt's rulers were believed to be the incarnation of Horus, the son of the sun god Ra. In China, the concept of Tian Ming legitimized the emperor, regardless of his social origins, as carrying the Mandate of Heaven. From Emperor Augustus in 12BCE to Henry VIII in 1534, heads of state have also been heads of their national religion. In charge of both the political machinery and the religious. Politics was charged with keeping order, increasing and distributing wealth, maintaining security. Religion guided leaders, but also kept the people in line. This created the perfect situation for the absolute ruler keen to be seen as chosen by God, intimate with God, answerable to God and, ultimately second only to God.

Although appointed and guided by God, rulers did not have the knowledge of religion (and its effect on the public) that the priests had and conflict would often emerge. In politics, it is often said that good people enter the profession with the best of intentions, only to be corrupted, worn down or distracted by the realities. Their altruism gives way to a careworn cynicism. That said, others join up because of the power and opportunity the job brings.

Much the same could be said of the priesthood. Some

were called by their god to administer, dedicate their lives, spread the word, and try to make the world godlier. In the Mayan empire, priests would offer their own blood up to the gods. Religious figures have put themselves through incredible agonies in order to commune with their gods and help others. Power is seductive though, and the power wielded by religion could corrupt just as much as the political or commercial power that it often found itself entangled with.

Hailing originally from a noble family of Aragon in southern Spain, the Borgias have gone down in history as notorious for their level of corruption, nepotism, influence and decadence. It all started with Alfons de Borja, a law professor who became a cardinal and, when no other suitable candidate could be found, Pope for three years. Alfons, then Pope Callixtus III, died in 1458 but not before he could appoint his nephew Rodrigo Borgia as a cardinal. Rodrigo in turn was elected Pope in 1492, becoming Alexander VI and it was Rodrigo who really worked to put the family on the historical map of infamy. He fathered a number of children via assorted affairs, including Giovanni (whom he appointed head of the Papal army) and Cesare (another cardinal). He married off his daughter Lucrezia to one of the most powerful Italian families, the Sforzas, and did likewise with another son, Gioffre, marrying him into the Aragon royal family. Giovanni was also used in a political marriage to stabilise relations between France and Spain.

It is tempting to view this assortment of nepotism and familial politicking as scandalous. By the standards of the time and of most aristocratic families in Europe and elsewhere, it was not. Like a lot of history, rumour mixes with fact. The malicious intent of later rulers driven to discredit past figures for their own ends can produce unreliable records. This could account for the long-believed stories of orgies in the Vatican hosted by Alexander VI and attended by his children.

Rodrigo/Alexander's children were, by most accounts, more of a problem. Cesare is said to have murdered Giovanni and acted ruthlessly in his attempts to manipulate Papal decisions and create his own Italian kingdom. Lucrezia, meanwhile, separated from her Sforza marriage, embarked on another relationship with a man later found dead (assumed by the hand of Cesare). Another marriage designed to assist Cesare's political ambitions also ended in murder. It is possibly due to the disreputable view of women in history that Lucrezia is also believed to have poisoned, spied, and seduced her own way to power alongside her brother.

Whilst revisionists question their notoriety, and new facts still come to light, the Borgia myth represents perhaps both the acme and the nadir of political, commercial and religious power and corruption all at the same time. They were not unique and the absolute power of the Roman Catholic Church in Europe from 1000CE onwards led to multiple examples of absolute corruption.

Religion and politics would remain intertwined for centuries to come, each lending power to the other. The second wave of Christian colonisation would come with the European colonisation of the Americas. Like their forebears the Romans, colonists would see their mission of bringing (or imposing) Christianity to new lands and new populations as divinely sanctioned. Those newly discovered lands also held great wealth, which was only apt reward for those that had risked everything to get there. Their perilous journeys made possible by the protection, guidance, even the will of God would bring earthly rewards.

This resulted in over two thirds of the world's population all believing in something very similar – namely the Abrahamic faiths. For the most part they were politically ruled by those professing to be appointed by, and doing the work of, the God they now all believed in.

Faith equating to notions of civilisation and morality held sway over many societies and justified what would later be seen as terrible, inhuman actions. We now see the error in forcing a population to give up their land, yield to the arbitrary will of outsiders, or be sold into slavery just because the conquering group believed they were somehow morally right. Yet faith remains vital to modern political leaders. Perhaps because religion became at least as much about culture, about holding a group of diverse people together, as about ideas of how the universe came about and was ordered. People fought for their god but in fact were fighting for a way of life, a human vision of the world.

Even today few political leaders either are, or would admit to being without a religious faith. The cynic would argue that it is simply a political necessity – most people profess to hold at least some religious belief so a leader seeking popular support needs to as well. People still see those with religious faith (or at least, a religious faith they are familiar with) as more honest; more likely to be 'good' and to do the right thing.

For those leaders that do genuinely hold a religious conviction, it is more likely to be a consolation in times of difficult decision-making than anything else. It is said that great power can bring great loneliness, and faith is a comfort. God and religion offers rulers a moral reassurance, that as long as what they do is right by their faith, they cannot go too far wrong. But faith can also be used as an excuse for a poor decision, giving a moral dimension and an honest, heartfelt motivation for a course of action.

In most parts of the modern world the religious hierarchy within a country still holds influence. Today, however, it is less to do with political rulers being advised on what God would sanction or what events a holy book might have foretold. Now it is more about the influence religious leaders continue to exert over parts of the population. If a senior religious figure takes the unusual step of condemning a political action, it can have real implications for a ruler or party of government. Even those people that do not share the religious leader's faith will still see their condemnation as carrying the weight of someone well-placed to judge what is morally right or wrong.

For some parts of the world religion and politics remain virtually as one; a country and its people cannot be ruled in any way other than the religious. Part of that stems from certain faiths inherently questioning the ability, or the right, of any human to rule over another other than by direct reference to God, or that only God can set down laws and modes of behaviour to govern humans.

The power of religion aligned with, and even exceeded the power of politics and monarchy in Europe from the 1100s onwards, resulting in a profound corruption of the Roman Catholic Church. An institution supposedly concerned only with God, faith and human morality was focused increasingly on money, politics and earthly influence. By the 1500s this had led to the Reformation and a generation of people asking fundamental questions about faith, humanity, God and the Church. By the 1700s the Enlightenment saw science and a host of new ideas fill some of the vacuum that a loss of faith had created.

Trade, the practical motivation for people to travel the world, underpinned the evolution and spread of religion in the first instance, and then of science and much else. Trade became as important to a ruler (secular or religious) as fortune in war and successful harvests. Just as those who understood war and good fortune, so those who understood trade became vital to power.

War and trade came to dominate the success of a nation state more than any other factor. Success in these closely related endeavours would have been thought of as at the gift of the gods. Despite making all the appropriate entreaties and sacrifices though, it became clear that success in war was frequently down to more pragmatic, human matters. The winner in a conflict would be the one who chose the battleground carefully and who had the best maps. Those that knew their enemies' resources, food stocks, and supply lines. Those that understood the size and number of their smithies or stables, and the type and quality of their weapons and fighters.

Around the 6th century BCE the Romans developed the art of the general; what was necessary for success in war. They applied a word derived from Greek, *strategia*. Strategy came to encompass an understanding of the enemy, and of your own assets, motives and abilities. Key to strategy was being able to predict what the enemy would do, and pre-empting or being ready to counter it. As well as access to funding and resources it was also necessary to understand the knowledge, training and personality of the opposing army and their leaders. A factual picture of the enemy, and as such the nature and scale of the enterprise you were undertaking, was vital to success in conflict.

The most valuable advice in such planning was human in

its nature rather than divine. Spies, cartographers, experts on weaponry, tactics and weather, those with insight into morale and the nature of a population. These were the advisors the successful state employed. Predicting how the opposition, as well as your own armies, would act and react were the decisive factors. In time, what was true in war would be seen as just as true in politics and commerce.

PART 2

ECONOMIC FORECASTING

Religion had built trust between nations, and where it was lacking, war saw populations conquered, cultures built, and new societies formed. Trade had cemented these connections and made them global. A globalised world, however, brought new demands.

Politics became increasingly reliant on trade and commerce, and the advantage in those areas went to those with the best plan – the more accurate prediction. Economics, the study of work, money and production, became a vital source of advice to political rulers. It also moved from being a philosophy, a set of competing political ideas, to being a science, a natural law. It came to supplant religion and mysticism as the preeminent predictive science.

Having built a reputation and a political and financial power-base, economics started to lose its predictive reputation through a series of ill-judged assumptions, corruptions and very public mistakes. This only came about, however, once it had become fundamental to how the world worked, and not just financially.

3

TRADE AND MONEY (THE NEW RELIGIONS)

Religion and spiritualism of one sort or another dominated almost every aspect of life for millennia. That meant those with insight into a world beyond the physical were revered. They were vital to the wellbeing and success of both a nation and its rulers. Their insights were rarely questioned, especially by the uneducated masses, and even by the powerful, for fear of the repercussions in this life, or the next.

As well as forming an understanding of the world and its mysteries, religion provided a less obvious use as a basis for trade. From the early days of the Agricultural Revolution, groups had traded what they had for what they did not have but others did. Those living on the coast would trade fish with those living in-land growing wheat or fruit. As seasons changed and areas offered up different crops or prey, so new people would be encountered, new trades made, new friendships and alliances sealed. They may only be met occasionally, but these social networks would become firmly established as populations grew more static and the need for trade as well as allies increased. Villages merged into settlements, towns and cities. Populations identified as citizen of somewhere, and those living in other places needed to be identified as friend or foe, useful or not, trustworthy or dishonest.

As a desire for trade brought populations into contact with new and increasingly remote groups, trust became a vital element of the equation. Individuals might know each other, but that was not enough; traders needed to know that they would not be cheated or that they would be safe inside the walls of another city. Religion with its sense of morality, laws and divine punishment formed the essential basis of trust; it was an identity beyond the geographical. Respect for shared laws, fear of the same spirits and deities, prevented traders from cheating each other. At the same time, beliefs and ideas about gods and nature were exported along with goods.

The desire, and sometimes necessity to trade also took populations overseas, in particular those living on small islands. Brave explorers journeyed many miles and found people quite different to themselves, who had developed along different lines. Those explorers would return home from making first contact with these new people with stories and goods. Word spread. New commodities, new products produced in new ways brought those wishing to purchase them from further afield.

Sometimes traders stayed in their newfound destinations, setting up markets, even colonies where they assimilated into the existing population. As trade grew in popularity and geographical scope, so too the exchange of ideas (including religion), which in turn led to innovation. The appropriation and adaption of ways of life, religion, language, improved methods of metalworking or agriculture, the results of collaboration all came from trade and the travel that it necessitated. Those that produced a commodity learned how to make them into more valuable products.

Communities were largely self-sufficient in the essentials of life. They would produce enough to eat, drink and protect themselves, otherwise they would have died out or moved on. Trade, however, brought a sense of value to things they did not need. You could live off the fruits of the soil in your particular area, but trade brought variety. Some items, like the fish from a neighbouring city were perhaps easily equated to, say, a bag of grain from an arable city. But what if acquiring some items required taking arduous, dangerous journeys? What if some products were simply less commonplace? Would that effect how desirable they were; their value to others?

THE INVENTION OF MONEY

Religion and shared or similar beliefs may have helped build and cement trust between disparate groups seeking to trade, but something else has proved to be a more durable, widespread embodiment of trust. But like religion it has also been the source of distrust, enmity, and the most brutal actions. It has inspired obsession and loathing, done great work and terrible harm.

Although it is widely thought that money simply replaced bartering because it was easier to carry around some pieces of paper or tokens than it was to carry a pig or a plot of land, in fact this may not be the case. What is more likely is that systems of credit and debit were established early on in human development, around 3000BCE. The early systems of record, which developed before writing, formed a way of tracking amounts of who owed what. Accounting for what was owned or being transferred in ownership was the purpose of some of the earliest forms of numbers and their recording. In his book *Debt: The First 5000 Years*, David Graeber argues that money did not replace the bartering system but pre-dated it. He suggests that the idea of being owed a unit of something emerged first, and that unit evolved into a physical, exchangeable thing, which in turn developed its own, discrete value.

As with anything owing its origins to a time before written records, this history of money is subject to debate. It may be the case that, as conventionally assumed, money was just more convenient, and that it allowed work or goods to be valued in discrete amounts. It facilitated the easy exchange of commodities both parties wanted. Otherwise the pig farmer was confined to only trading directly with those that wanted his pigs regardless of whether he wanted their bread or their masonry skills. The credit and debit theory also accounts for a world before money in which some would be seeking to derive worth from something that cannot be easily divided or would only be wanted in the future. So the pig farmer could trade his pig not for 50 loaves all in one go, most of which would go off before he could eat them, but 50 loaves spread over the course of time. Or, were he not also a butcher, he would not have to cut a chunk of his pig off to exchange for a day's labour from a farmhand. Equally, if our pig farming friend did not want any bread, but wanted wood for a fire or wanted someone to build a wall to keep his pigs within, what happened to the baker keen on some pork for his supper? And what if he wanted apples but a bad harvest that year (perhaps due to a particularly vengeful god of wind) meant there were fewer around? Or if another pig farmer with a smaller family to feed was offering the same size pig for fewer loaves or hours of labour? It is easy to see that, human ingenuity being what it is, a solution to these complex exchanges would quickly emerge. Something other than just an agreement between parties, some of whom may live many miles away or be met only very occasionally.

Somewhere along the line a solution to the problem of people exchanging and owing goods and services indirectly became a matter of necessity. The vital question was, how to account not just for how much someone wanted something, but also its scarcity, quality, the effort involved in producing or growing it, the time of year or events from droughts to mining disasters? Money, little more than an agreed upon number, emerged as the solution.

The physical nature of money, as a thing that represented the value of something different but with no literal value itself, has been varied, from stones and pebbles to animal teeth. Currency as we would recognise it is thought to date back to around 600BCE (although in China, around 1100BCE, they are thought to have used small bronze sculptures of the item being bought or sold).

Most people would think of early money as metal tokens; coins. The best-known metals used for currency being gold and silver. These two metals came to form the basis of global currencies through an essentially unique (in a historical context) combination of durability, flexibility, size, accessibility, rarity and appearance. Gold in particular fulfilled all of these criteria. It did not tarnish or oxidize so it lasted, even when exposed to harsh conditions. It could be worked into a wide variety of shapes and even sculpted making it perfect for decoration but also coinage. It could be hammered thinner than paper or formed into bricks and could last centuries for those seeking to accumulate lots of it. It was just the right level of rarity – common enough to be mined effectively, but hard enough to find to make the supply small yet fairly consistent (with, perhaps one exceptional point in the 17th century). Perhaps above all else, the basis of gold's early worth came from its sheen, and the permanence of that sheen. It could well have been that quality which meant gold was valued in early civilisations for its similarly to the sun, raising it to the level of a religious, near-godly entity.

This adoration was certainly true amongst the Aztec Empire of South America. They used gold extensively, but only when the Spanish conquistadors of the 1600s arrived did they come to understand the terrible value it had had bestowed on it as a means of wealth. The Europeans inflicted huge suffering on the Aztecs, enslaving them and effectively wiping them out, all because of the value gold had acquired in Europe and Asia.

Physically, money came to fall into two categories: that considered of value in itself (predominantly gold and silver, but also bronze and sometimes iron or copper), and money that it was agreed represented an amount of debt, most obviously paper money but equally coins of more base metals or alloys. Tokens were used to represent a holding of gold – a debt that could be called in if required – from someone trusted to hold that which has real value, usually protected from being stolen or tampered with (generally a ruler or a bank).

Currency, that is money that is being circulated, became the foundation of states and the barometer of their success. Holding a store of valuables which backed or underpinned a currency became the job first of local feudal leaders, then of banks who were (theoretically at least) independent of political leaders. The amount held by a particular city or state was the cause of great rivalries.

It came to be the case that the most important currency was that which had the most value to the most people, regardless of who they were or where they lived. That, of course, would be gold and silver. They became the international currency of trade and exchange; that against which all other currencies were measured.

The accumulation of gold in particular became the principle preoccupation for rulers and states. Possession of gold became a measure of success and security for the future. Its global movement would, over time, build and destroy empires, kill and maim millions (probably billions), and push humanity to incredible levels of achievement and desperation. Its collection and protection, first through war, then through trade (and more often both) would move rulers to rely less on generals and more on merchants, bankers, and later those who studied the production and movement of money; economists.

INTERNATIONAL TRADE

Having found a way to trade effectively with trusted partners, in effective markets, only one barrier to development remained. Fundamentally trade occurs when there is a geographic separation between production and consumption. Although it would, in times of scarcity, be necessary to trade, few ancient populations would rely exclusively on something they did not produce themselves. Trade often started out as a luxury, for the wealthy and privileged. As a state increased its wealth and that of its population, those luxuries would become ever more accessible to more and more people. Demand, and the supply to meet that demand, would dictate trade and prices. This had a profound effect on a population, changing tastes, necessitating new skills, pushing certain individuals to take great risks to find new ways to meet demand.

In the first century CE large Roman garrisons insisted on some home comforts as compensation for being stationed in the bleak north of Britain. The importing of their preferred produce left its mark forever. Wine, pork and oils from the Mediterranean replaced the mutton, mead and misery of this newly conquered (or in some cases stubbornly unconquered) land. In time this changed tastes and established a market; wealthy people willing to pay for products they did not need but wanted. With enough demand, and enough money to pay, the difficulty and cost in transporting items hundreds of miles became worthwhile.

This also demonstrates the impact of specialisation, something that would go on to shape states and populations over the centuries. With money and trade came the ability for certain individuals, and then entire settlements, to focus on doing one or two things really well rather than trying to be self-sufficient. Why produce really high-quality leather goods but only average meat when you could instead focus more of your efforts and resources on the former, improving and producing more. You could then sell more of your leather for more money, and then buy more and better meat from a village that has the land and expertise to specialise in that?

The complexity of trade gives some indication why some believe it to be the single greatest achievement of humanity alongside language, writing and the making of tools. It has had such a wide-ranging and fundamental impact. It took centuries to realise, and, like language, evolved often independently amongst populations that had not intermingled for millennia.

Trade became the principal influence over the development of local economies, buying and selling between towns and villages. For all its obvious advantages and uses, however, trade on an international level remained limited in its wider influence. Although some cities thrived as hubs of wider trade, the wealth of nations depended principally on the acquisition and protection of land and the sustenance of its peoples. Through war and treaty, people, natural resources, treasure and crops were brought under a single ruler. Thus empires were constructed.

As the populations of towns and cities grew, there was a greater demand for food and shelter within a small area. But more people also brought greater production of tools, food, goods, and ultimately wealth. It brought larger armies and provided them with more and better weapons. It developed new skills and new inventions.

Yet the economic and social order remained relatively simple. People largely did as priests and the representatives of the ruling classes ordered them. The success of any enterprise depended simply on human endeavour, the weather and good fortune (as controlled by the gods and understood by those that spoke for them or interpreted their whims). The lives of most people were largely immune to the actions of other nations and populations thousands of miles away.

In trade, Ancient China was something of rarity, especially for its size. Its provinces and regions, more usually rivals or even outright enemies, were finally (more or less) unified around 220BCE into what would be seen as a centrally governed country with borders similar to modern China. Shortly afterwards the ruling Han dynasty established the so-called Silk Road, a trade route running through Asia and Arabia to Europe. A lucrative trade in silk and horses ran along this path with wealthy elites throughout these regions paying well. The nascent Chinese state grew rich and influential. The wealth derived from this international trade enabled China to fortify itself and subsume smaller kingdoms.

Trade along the Road expanded with ceramics and foodstuffs, and China became the leading trading nation of the world through most of the Middle Ages. Despite wars and internal strife it maintained that status. Its wealth enabled new innovations that served to further secure its place as a hub of trade and a magnet to people from all over Asia and Europe. This mix of people in turn served to create a virtuous cycle of income growth and further innovation.

Meanwhile in Europe, a continent slowly but increasingly comfortable with trading overseas, the resultant population shift towards cities and towns combined with trade in one particularly devastating event, or series of events; the Black Death. As larger and larger groups of people moved closer and closer to each other, first in villages, and then towns and cities, so disease spread much more quickly, affecting many more people. At the start of the Agricultural Revolution the threat was from diseases originating from newly domesticated livestock. With the growth of trade, ports in particular were open to new diseases from abroad. A large, condensed population were exposed to diseases they had no immunity to and no understanding of how to treat. In just one of many examples of complex, unintended or unforeseeable consequences of trade, it would also provide a turning point in who would dominate global trade.

By the mid-1300s a feudal system was well entrenched in much of Europe. A landed elite owned most of the productive agricultural land. Uneducated, semi-skilled peasants made up much of the population and they lived on the land at the munificence of powerful landlords. They swore loyalty to the landowners who set and enforced laws and taxes at the behest of the monarch. At the same time the Persian and Chinese Empires were much more advanced and diverse; they had greater cities, greater learning, and dealt more in international trade, finance, ideas and art.

Climate change in remote parts of northern Asia killed crops and pushed rodents into more populated areas. With them came fleas carrying bubonic plague. The plague travelled along the Silk Road around the mid-1300s devastating the Persian Middle East as well as parts of Europe as it went. Naturally centred on trading cities and ports, the epidemic primarily affected urban areas, leaving rural communities relatively unscathed. Despite the death of tens of millions across Asia and Europe the population densities meant the more rural, less advanced Europe was less affected and could rebuild more rapidly in the wake of the plague.

The aftermath of the Black Death saw European populations move to repopulate the cities. The killing off of a lot of people meant fewer workers, which meant higher wages for those that survived, whether rural or urban. This shifted somewhat the balance of power (albeit temporarily and in a limited manner) away from the landed classes, who in those times wielded political as well as social and commercial power. This shift weakened the powerful elites, and gave some of the working classes disposable income for the first time. This meant a new market for imported goods, and a healthier, better fed, more productive population.

A fatal plague realised by weather conditions and spread by trade had decimated the trading nations of the Middle East but set the lesser-trading nations of Europe on a path which would ultimately lead them to dominance.

It is not easy to rule a country as large as China, and many Chinese emperors were ruthless and authoritarian, aided by a bureaucratic culture born of the rationalist philosophy of Confucianism. In the late 1300s, Emperor Zhu Yuanzhang, founder of the Ming dynasty, sought to enforce a model of rural self-sufficiency and a rigid military structure throughout the nation. This included significant improvements to the navy. As part of his revolution he also broke down the existing and, as he saw it, entrenched bureaucratic structures. He deployed his relations throughout the country to rule, giving each a personal army, but imposing rule centrally.

Zhu Yuanzhang's successor, Zhu Yunwen, attempted to undo some of these regional structures leading ultimately to a revolt and his overthrow. Zhu Di, the third Ming Emperor restored order, a part of which meant rebuilding the bureaucracy around, amongst others, a class of Confucian scholar-bureaucrats, including one, Zheng He. With decades of investment and innovation in shipbuilding to call on, an adventurous and influential Zheng led explorations of the oceans around China. He travelled to Arabia and East Africa, even, it is thought, to Australia. Each ship in his fleet incredibly advanced, fast, well-built, and carrying thousands of crew. A formidable Chinese armada of exploration travelled far and wide.

Chinese power, ideas and culture were exported all over the eastern world establishing the country's primacy in the region. However, internal disputes, shifting priorities and conflicts with neighbouring kingdoms saw the bureaucrats judge expeditions like Zheng's a waste of treasure and manpower (some of which was diverted to reinforcing the Great Wall). By the 1430s China had all but stopped their overseas expansion and the navy fell into disuse. Shipbuilding and seafaring skills waned and China looked inwards.

The prevailing opinion was that this mattered little. China had contacted many of the ruling elites overseas, in some cases the first foreigners ever to do so, and their place as the local superpower was firmly established. Other nations came to them to trade goods; expeditions to far-off lands were now an expensive and unnecessary indulgence.

By now, 6,000 miles away in Europe, the Portuguese and Italians were strengthening their sea legs and were starting to sail across the oceans. They started to expand their powerbases, and more important than any cultural influence or reputation, their trade routes. Perhaps, had the Chinese known that the Americas lay the same distance to their east as Europe did to their west, they might not have been so quick to withdraw from international exploration. But withdraw they did and the effects of that decision would resonate for the next seven centuries and beyond.

Following the paths first cut by ancient Greeks and Romans, and later by the likes of Marco Polo, Europeans started to turn up in increasing numbers on the shores of China seeking to trade. By the 16th century European trade routes over land and sea were firmly established, and well protected, both to the east and to the west.

The links to China were prized for the supply of spices, silks and foods and the Chinese economy started to flourish again. With few European products that the Chinese wanted, trade relied largely on silver, which fundamentally changed their economy. With the country increasingly reliant on selling and exporting its goods, a combination of crop failures and natural disasters (and a failure of priests and astrologers to foresee such bad fortune and weather) saw a collapse which brought about the end of the Ming dynasty. An entire royal household brought down by trade, poor planning, and a policy of insularity.

China's withdrawal from the world, in the circumstances quite sensible, would go on to affect it for centuries to come. It would arguably miss out on becoming the pre-eminent global economic power (all the necessary elements were in place ahead of most European rivals). Instead it would become a pawn in the trade wars of other nations.

4

TRADE MEANS BUSINESS

Those living in port cities across Europe, Asia and parts of Africa saw trade every day. Their local economies were sustained by the presence of goods passing through, and those seeking to buy and sell. Foreign languages, cultures and ideas mingled with the domestic and they became hubs of innovation, novelty and stories. They also attracted the wealthy and those seeking wealth (by whatever means).

Foremost amongst the wealthy were the merchants, whose days were filled with attempts to find new goods to trade in, from new sources, to new markets and destinations. They sought to buy for less and sell for more. They wanted to ship quicker and cheaper, and to avoid any disasters that might befall their goods - usually storms on the sea or war on land (or theft on either). As such, assessing and managing risk became vital parts of the successful merchants' expertise.

As with most ventures, with risk comes profit. The greater the risk, the greater the costs, but the greater the risk the more you could charge for your goods, and the fewer competitors. International trade was expensive, especially by sea. Hiring a ship (preferably a good one), crewing it (with a reliable company under a capable captain) and supplying it was a costly business.

With luck the ship would safely deliver its cargo, receive prompt payment, and return with the money (and some more, valuable goods), out of which the costs could be paid and a healthy profit made. That profit, however, also had to account for the losses made on similar journeys. It had to compensate for the storm-wrecked ships, the pirated cargos. These journeys to and from market took months, sometimes even years, so it became necessary to have a flow of money and means to pay some costs in advance. Merchants also sought to spread their risk; bringing in investors and backers from elsewhere instead of staking all of their own money. So with the rise of trade came the rise of banking.

Banking had been around almost as long as money and originally served as a source of credit, and for the exchange of currencies. The growth in trade from the 11th century saw the rise of the merchant bank. Concerned primarily with commercial loans and investments, the uncertain nature of international trade introduced an element of risk and therefore speculation to their work.

Bankers would have to assess the risk against the reward of putting money into a venture seeking, perhaps, a new market or source of goods. Both merchant and backer would judge what they could afford to lose, over how many journeys, and the chances of a successful trip compensating them. They would consider the route taken (a quicker route, but a more dangerous one, versus a better-known, safer route) and the season (information on weather patterns being limited in those days). This risk also inspired developments in insurance.

With every uncertainty comes some attempt to predict, usually by those who think they can steal a march on competitors. The nature of the risk, the likelihood of success, the cost and degree of failure; these were all factors that needed to be understood. It is safe to assume that in the early days of these ventures those involved were little more than gamblers, working with instinct born of hard-earned experience. A rough feeling for the plausibility of those involved, a knack for spotting a worthwhile idea, and possibly some knowledge of elements like the cost of the goods to be shipped and the ability of the ship's captain and crew.

Humans do not like uncertainty, whether in life or in money. Bankers and merchants were particularly keen on removing uncertainty; on investing rather than gambling. That meant assessing, measuring and putting values alongside their endeavours. Numbers started to be the key to understanding risk and rewards; analyse the numbers scrupulously enough and you will know how much money you stand to gain or lose.

Most important in these investment decisions was the cost and value of the goods involved by the time they reached their destination. Whilst the likelihood of a cargo actually reaching its destination was hard to know for sure, experience suggested it happened more often than not (otherwise you were not a very successful trader). As long as the successes outnumbered the failures, the failures could be mitigated, ideally by not staking too much on any single enterprise.

Trade developed to a point where all reasonable attempts to mitigate what could not be measured had been covered – the ships, the weather, the type of cargo, its security. The market value of a cargo then became the key factor in investment decisions. Who else was due to bring in a cargo of pepper this month? What effect would the loss of someone else's shipment of silk have? Would war with a neighbouring country make tea hard to come by? These supply and demand issues dictated the price that could be asked at market and as such the profitability of a venture. But these aspects were measurable too and, over time, market prices came to underpin the economies of trading cities.

As well as market prices and wealthy merchants, what influenced port cities the most was diversity. A constant flow of people originating from or having visited places and cultures thousands of miles away. They would bring the influence of different beliefs, new perspectives, and of other languages. The greatest, most enduring effect of trade would be not in goods but in ideas – ideas combined, viewed differently, incorporated, worked on and developed. The market came to appreciate this melting pot of humanity and ideas, but it struggled to value it. It could not say whether an enterprise was worth investing in, just because it might, indirectly, bring about new ideas at some point.

TRADE AS NATIONAL IDENTITY

Trade of this sort carried on largely unaltered for a few hundred years. Cities grew rich and powerful, but the majority of a nation's people lived in rural parts, untroubled by the chaos of ports and markets. It took until the 16th and 17th centuries for trade to really start to establish new power structures and to shape the modern world. Having witnessed the faltering of China in the global trade story, the focus falls to Europe, in particular western and southern Europe - Britain, Spain, France, Portugal, Holland and Italy.

Despite its modern-day reputation, the small island of Britain was not always the seafaring superpower its people would have the world believe. The Tudor monarchy had renounced Roman Catholicism when Henry VIII felt his loyalty to the Church had not been repaid and he tired of being subject to Papal decree. His youngest daughter, Elizabeth, aimed to carry on where her father had left off in building a country and a monarchy of iron resolve combined with rule by wisdom and reason.

With a backdrop of papal support and personal resentment (after England had supported his Dutch enemies), Philip II of Spain planned to invade England and re-establish Catholicism as the religion of both the nation and the monarchy. The key weapon in this invasion was the Spanish Armada of 1588. Facing Spain's well-equipped navy the only feasible option for the English was outnumber them. Drawing together all the resources it could, an English fleet augmented by private vessels and crew, including privateers (essentially seafaring mercenaries) Sir Francis Drake, Lord Howard of Effingham and Sir John Hawkins, met the Armada.

Through a combination of strategy, persistence, weather and good fortune, England triumphed. The result, apart from the continued reign of Elizabeth I and the dominance of Protestantism, was an increased investment in naval technology, weaponry, training and weather measurement. It was a narrow victory though, and the Armada emphasised Britain's susceptibility to naval invasion. Next time they might not be so fortunate. Considering offense as the best defence, England undertook greater naval expeditions, with new, improved ships and better-trained officers and crew. Naval colleges were set up, money and men were poured into developing seafaring and navigational technology. It would be the dawn of Britain's reputation of ruling the waves.

Foremost amongst the early targets of English expeditionary attempts was the Barbary Coast, now the North African coast, initially picked out as a provocation to nearby Spain. Through ambassadorial visits and Elizabeth's personal letters to Ottoman ruler Sultan Murad III a partnership was sealed. By viewing the sultan as a fellow divinely chosen ruler, and by underlining the similarities between Protestantism and Islam, Elizabeth used religion and royalty to create a foundation of trust and trade and a major victory for the English.

Whilst the potential national interest in exploring territories far beyond Europe was clear, treasury funds were not limitless. Expeditions around Africa and Asia were high risk and very expensive, regardless of the great rewards promised. Private wealth in the shape of merchants and their banking partners saw an opportunity. They could both further national reputation and wealth whilst also establishing new sources of raw materials and new trade routes. Although carried out in the name of the nation state, wealthy individuals took an interest in this new type of exploration and the speculation attached to it.

Undertaking any enterprise in the name of the state required royal patronage. A group of merchants banded together to petition Elizabeth I for a charter. It would grant exclusive rights to continue their individual, nascent endeavours exploring the East Indies, specifically the region now known as Myanmar and Thailand. A company of merchants and bankers would usually only exist for the duration of one expedition, formed to fund, plan and exploit just that single return journey. Merchants combined funds, expertise, and goals to lower their risk. That journey complete, the company would dissolve and they would move on to the next project. This project, however, would have a longer-standing outcome.

The Royal Charter granting exclusivity over a trading route to the East Indies for over a decade was approved and this company of traders would become the Honourable East India Company. Focusing primarily on the spice trade, the East India Company (EIC) would become the most infamous of the overseas trading enterprises. It was the first private company in modern terms, and it wielded a monopoly over this lucrative trade route. It was a monopoly the Company would guard ruthlessly and that would influence future trading endeavours.

Across the North Sea, the Dutch East India Company (or VOC - *Verenigde Oost-Indische Compagnie*) was established along similar lines to their English rivals. Formed from national political aims, through military necessity, and, of course, financial ambition, the VOC saw what the English had done and moved quickly to compete. Both companies would expand across the globe from America and the Caribbean to the Philippines. They would both be granted the rights to raise their own military forces. They would seize competitor's vessels and their cargo. They would forcefully subdue local populations. They would fight trade wars, impose taxes and execute their own laws. They would become as powerful as any nation state, but without any of the duties or responsibilities, apart from to their owners.

The Dutch and English companies were joined by other European companies from other nations in an increasingly competitive, and brutal race. Despite their size and power, the trading companies could not escape the rule of the market. The cost of gathering, shipping and protecting pepper, silk, cinnamon, coffee and the like would still fluctuate, and so did the profits. Attempts to predict the market price, supply and demand, what competitors were able to charge, were limited. However, with exclusive rights to shipping these monopolistic companies were able to manipulate these factors. The companies had the wealth, power and control over supply that meant they could affect markets by warehousing goods until their value rose, or by blocking competitors' attempts to ship the same goods.

As trade opened up around the world from Asia to the Americas and into Africa, so the trading companies grew in wealth and power. The VOC became the first multinational company. It publicly issued bonds (essentially portions of company debt) and shares, becoming the first publicly listed company in the modern sense. Allowing the public to own it and lend it money gave it even greater social and political power and influence. It was the first company to acquire quasi-governmental power, enabling it to enforce its own laws with imprisonment and execution, to wage war, issue currencies and negotiate treaties.

Beyond these powers, the VOC also, deliberately and by accident, initiated many other modern corporate fundamentals. It created what might be considered a brand identity. It established the typical operation and hierarchy of a corporation. It invested and diversified, including investing directly in overseas projects and infrastructure. Even more progressive ideas such as business ethics and culture also have origins with the VOC. It helped make the Netherlands one of the wealthiest nations on the planet and in doing so cemented the relationship between government and privately financed enterprise. The VOC enriched both the nation state and individuals, paying shareholders over 10% of their investment every year for almost 200 years. It employed thousands, engendered great national pride, increased the overall standard of living in the Netherlands, invested in public institutions, and paid substantial amounts in tax.

For two centuries the trading companies continued along these lines. Finding new products to import and export, discovering new lands and markets, and new routes. They would engage in wars and find new sources of profit. As time went on the nature of global trade settled once again into some sort of equilibrium. It was more influential now, but still, it is estimated, only accounting for around 5% of the GDP of most European nations. It was difficult and risky, and served only a relatively small elite (albeit an incredibly powerful one) of merchants and wealthy markets. With a relatively small market to deal with, trade improved only by degrees. Competing companies invested in better ships and crew. They bought more influence over national rulers and the laws they set. Bigger breakthroughs, however, lay in more effective militia, improved navigation and weather forecasting.

5

THE SCIENCE OF WAR AND WEATHER

When the world works in ways humans cannot possibly comprehend, they come up with something to fill the gap in their knowledge. 'Why did this event happen?' soon becomes 'how could we have foreseen it?' The question of 'what happened?' becomes 'what conditions existed leading up to it?' Belief and early science were partners in attempts to find explanations. To modern minds the early theories may seem ludicrous, but a great deal of time, effort and learning were dedicated to furthering the understanding of astrology, alchemy and so on. Whilst these pseudo-sciences lay the foundations for what would be considered true science, at the time much of it was focused on predicting events.

Although attempts to understand the natural world are as old as human cognition, around the 16th century science found a revolutionary zeal. It was the dawn of empirical science. A broad set of rules for analysis and understanding; of formulation and prediction. It questioned, theorised, disproved and, critically, by doing so, it progressed.

With science's role as a source of knowledge and understanding came an important social influence. Before science started to reveal new facts about the world, academic study and education was focused on maintaining the established order - be that regal, religious or military. Science meant challenging the assumptions of the past. It meant questioning power and even God. This was dangerous. Science had to confine itself within certain parameters for many years. Even then many scientific thinkers would be banished, condemned, even killed, or at least see their work destroyed or discredited, just for challenging the old ways.

Study, especially scientific study with its need of equipment and time and repeated experimentation, was expensive. Only the elite had access to the resources to fund science. As a result their influence over the work of science could be significant. The wise scientist would balance their work with the priorities of those in charge.

Before humans had written records, before they gave up the nomadic hunter-gatherer life, they engaged in conflict. The motives changed, the methods evolved, but the act of one group uniting and taking on another in physical combat has been a constant means of asserting authority over others for thousands of years. Whilst other, subtler ways to expand or secure power and wealth based on trust and agreement were commonplace, war has been a constant concern for rulers.

In ancient times, if the land on which your tribe had survived suddenly turned unproductive you could seek out somewhere else, somewhere that could sustain life. But there is a good chance someone got there before you, or thought they had some greater right to that land than you did. They might even think that they had a divine right to that land. It might be that two groups could share the land, creating a new, bigger tribe; if the land could sustain that many people and the two tribes were similar and trusted each other. Alternatively they could fight for the land.

Fighting for survival gave way to fighting because you were told to. Throughout history, young men have been sacrificed on the orders of older, more powerful men. The way in which men were persuaded or coerced into fighting varied, and over time became more sophisticated. Charismatic leaders would frame their motivation for war as beneficial to all of their people. They would fight for the glory of the state or the improvement of their lot. They would fight to defend their state, their reputation, and the 'purity' of their nation and the safety of their families. The real motives could be less convincing.

War would, for the most part, ultimately be waged for three main reasons – imperial, financial and religious. In the first, powerful rulers wanted more power; to control greater populations, secure their reputation and establish a legacy. Religious wars were usually inspired by those compelled by their god to convert disbelievers and heretics or to take back holy properties held by others. In time the definition of religion in the context of war would expand to take in the ideological as well as the spiritual. Financial wars were usually wars of trade, expanding and securing trade routes, appropriating the wealth of another nation, and securing new sources of raw materials, whether that is food or gold.

The purely imperial and religious motivations, seen in wars of conquest from the Akkadian Empire (in the 24th century

BCE) to the Napoleonic via the Roman (which served both goals at various points), waned throughout the second millennium CE. It became clear that war, in many respects, was just a more costly version of trade. The access to wealth in particular, could often be more cheaply acquired through trade and treaty. On occasion one necessitated the other but ultimately wealth was the justification for war. War was expensive, both in treasure and reputation, so would tend to be avoided in modern times. Unfortunately, whilst the number of wars decreased, the brutality increased.

Empire building, however, did provide a new impetus for science. Europe's imperial dominance of the world came about in large part because it combined the search for wealth and power with a search for knowledge. That search fuelled scientific understanding, technology, and in turn military power. Just as science works from the basis that it does not know and so seeks to learn, so too modern European imperialism. By acknowledging there were lands and people they did not understand but should, empires expanded along with knowledge. That is not to say both scientist and imperialist were not arrogant and destructive in their pursuit of knowledge. It is just that other empires, Roman, Han, or Arabic, sought power, wealth and religious expansion over knowledge and ultimately paid the price.

The discovery of America confirmed European supremacy. It is hard to underestimate the effect of discovering a whole new continent. Something so fundamental that questioned centuries of observations and assumptions scientific, political and religious. Prior to the late 1400s the populations of Europe, Asia and Africa believed they were the world. Finding America meant conquering it, which meant a European race to gain a foothold; a competition between leading European powers. It led to investment and advances in transport, tools, scientific understanding, the exploitation of new resources, new collaborations and enmities. More than that, it meant being able to accept that what was once certain was actually wrong. Not easy for a species used to being in charge, but vital to its progress.

The battle for dominance over trade, whether in the Americas, Asia or beyond, meant war. Success in war has always depended on a number of factors. Some were in the control of the rulers that declared war and the soldiers that carried it out. Training, weaponry, manpower, motivation, strategy could all be superior to the opposition. The wise ruler would not declare war on a state they new to be better at any of these.

Finance also played a key role. Paying, equipping and feeding soldiers would cost, especially if the theatre of war was overseas, and it may be that funds would dry up whilst workers fought and trade routes were blocked. Careful diversion of wealth to the running of a war was down to the ruler and the banks. Some of the first forms of taxation were raised to fund wars - with the intention that success in the war would bring more wealth to pay for the debt accumulated in waging it.

Chance, luck or fortune would also be vital to success in war, or would be blamed for failure, usually as a scapegoat for other shortcomings. A good army could be defeated by a lucky (or unlucky, depending on your perspective) shot taking out a brilliant general. In this, God and superstition would retain a role in a martial success. Weather would also be decisive factor. Overwhelming superiority in numbers or equipment could be quickly overturned by an inability to send supplies because of adverse weather or by ships being blown off course.

For a long time fortune and weather were viewed much the same. In order to win favour with the gods of weather and war, a Mayan or Roman priest might order sacrifices to be made by the rulers in order that they might triumph. But few rulers would rely solely on divine caprices. Those tribes or nations seeking to expand their territory learned that victory went to the best equipped, best prepared side. Excellence at war was a much-envied and decisive attribute.

Even before science was really science, the greatest minds in a nation would be turned to helping their country triumph in war, and thereby become a great nation, a wealthy nation, under a great leader. The weapons used to hunt animals were applied to conflict. Innovations in weapons made for more efficient, effective hunting and provided an advantage in combat.

Most advances in war, the elements that gave triumphant empires and nations the leading edge, came about through brilliant tacticians. For centuries wars were fought in much the same way. The technology of war, the weapons and supplies, advanced through two means. The first and by far most dominant was through gradual improvements from artisans (the makers of weapons) and soldiers (the users). The second, and for a long time a secondary method, was via scientific advances.

Although discovered in China around 1000CE, gunpowder did not find a military application for another 200 years. It was discovered by accident. Metallurgy had many more applications than seeking stronger, longer-lasting alloys for blades and armour and was primarily the concern of smiths and metalworkers. Understanding forces like tension and gravity applied to catapults was more about trial and error.

Science, in its earliest incarnation, was about education

and learning; about maintaining the status quo between rulers and ruled. Whilst social order and simple human curiosity drove scientific advancement, the cost and resources required for research required rich, powerful sponsors. Those sponsors sought to derive some gain from their investment. Some would hope for a purely commercial advantage. However developments from explosives to manned flight to nuclear technology saw other application. The winning of wars, regardless of the scientists' original aim, persuaded rulers to dedicate personnel, money and resource to scientific method. In order to progress, science needed the funding and support of government and commerce. As such science has frequently reflected the priorities of its patrons.

Captain James Cook's first voyage to the South Pacific was ostensibly a scientific mission. It carried a variety of scientists and gathered samples and observations from many lands. When it returned to England, its work vastly furthered scientific understanding. Cook's ship, the HMS Endeavour, was a Royal Navy ship and it carried a military contingent. Whilst at sea Cook claimed islands including Australia for Britain. They found new sources of trade and wealth, new routes, and even played a role in defeating scurvy, the disease that claimed half of all ships' crews at the time. In an age where information was slow to filter through to competing nations this innovation alone greatly aided British naval dominance for years to come.

War would become a leading (but by no means sole) motivation for scientific achievement and the technology it inspired. Throughout history, new, more devastating weapons have been invented. Sometimes by accident; a discovery that was supposed to benefit the world was applied in a way the inventor did not predict. Sometimes inventors believed that their clearly destructive invention was so lethal that no one would ever dare oppose it, effectively putting an end to armed conflict.

Richard Gatling invented agricultural machines before developing the pre-cursor to the automatic machine gun that bore his name. He believed that his weapon, the first to be able to fire hundreds of shots in rapid succession, would mean the end of large armies dying in open combat. If both sides had one, he reasoned, it would just mean random slaughter to no ultimate gain. It would render war futile and encourage other methods to settle disputes. Mikhail Kalashnikov, the man behind the AK-47, similarly regretted the widespread availability of the weapon that carried his name believing (perhaps naively) it was a tool of defense, not attack. Most infamously, many of the brilliant minds involved in the Manhattan Project and its successors had initially worked on atomic energy for peace or to simply further understanding of the physical world. They saw their work shift from scientific exploration and theory into an arms race; first with Germany and later with the USSR. Rather than cheap energy their work created the ultimate weapon, capable of destroying entire populations - although it is worth noting that some scientists did join up in order to help beat the enemy to the achievement.

With war and trade closely related it is not surprising that, just as science was applied to gain advantage in war, so it would be applied to trade. Whilst the commercial world embraced science in the production of new goods from raw materials, or the application of new methods to produce existing products, the nature of trade itself also looked to science to gain a competitive superiority. Where ships were concerned, the advantage in both war and trade went to the fastest, most robust vessels. Science played its part in new shipbuilding methods and materials, the use and design of sails, and early ideas of around aerodynamics. It also made improvements in the loading and unloading of cargo, the defence and security of vessels, and cartography. Being better than the competition in these areas all provided a small advantage.

Science had made marginal improvements in many areas affecting trade. Martial strength could protect against or prevent conflict or theft of cargo. Ships were quicker and more robust. Maps were more accurate. But prayer and sacrifice had done little to mitigate the impact of weather on shipping. The greatest threat to men, ships and cargos was sailing into a storm or being blown off course. Being able to understand the nature and direction of weather would provide a huge advantage.

Of all the inexplicable vagaries of the world, one of the most important to daily life was weather. Weather has always been vital to the wellbeing of a population, never more so that when shelter was basic and harvests fragile. Harsh or unexpected, unplanned-for weather could destroy a village or town, even a country. Weather was a mystery, and despite better housing and planning continued to wield great power. Even during times of expansive trading, a region still depended on its own ability to produce enough food within or near its borders. A drought or flood could starve a population and cripple an economy. For the elite, merchants and banks staked large amounts on enterprises that could be wrecked completely by the weather.

Weather is so important that huge amounts of effort, money and intellect has been dedicated to trying to understand it and predict its movement. For millennia weather was so hard to measure, analyse and predict that long-held meteorological superstitions remain with us today from groundhogs to the colour of the night sky to the reaction of seaweed. The desperate importance of predicting the weather, and the efforts involved in doing so, resulted in great scientific advances. In ancient times, however, weather remained a mysterious act, one so powerful that it could only be governed by something beyond human comprehension.

As fundamentally important as weather was it was incredibly hard to understand. It was everywhere but invisible; it came from nowhere, disappeared, moved, but could leave devastation in its wake. No human could affect it, which is one of the reasons it was held to be solely the work of the gods and spirits. The sun was central to scores of systems of belief all over the ancient world, hailed as the giver of life. There were gods of wind and rain and thunder; powerful gods controlling powerful elements upon which lives depended. Each god would demand sacrifices, prayer or obedience in order to intercede in human affairs, deliver good harvests, punish or destroy enemies and their cities, and influence rulers.

It is suggested that in the Battle of Orleans, Joan of Arc influenced the wind though prayer, enabling ships to enter the besieged city and deliver provisions. The Spanish Armada's attack on England was hampered by unfavourable winds - no doubt thanks to God favouring the Protestant English or possibly part of a bigger, more complex plan for Catholic Spain. Monsoons prevented Mongol leader Kublai Khan's imperial expansion into Japan (after which the Japanese Shinto priests coined the term '*kamikaze*', or divine wind). Napoleon's march on Russia was famously impeded by inclement weather. The French Revolution was precipitated by the joint economic effects of war in the Americas, a drought, and a crop-destroying storm. If only those in charge at the time could have seen these events coming.

Around 480CE, the powerful Persian Empire threatened the very existence of Greece. Military leader Themistocles 'read' the winds and tides at the naval Battle of Salamis and defeated the Persians. Around a century later Aristotle wrote the first known study of weather. Although much of his work would look simplistic or just plain wrong today, his analysis provided the foundations of later experimentation and understanding for meteorological studies. It was translated into Arabic and Latin and influenced thought on weather and physics until the Renaissance. Touching on geography, astronomy and chemistry, he defined the elements of fire, air, water and earth. He attempted to understand the conditions under which clouds would produce lightning. Whether scientific or an attempt to understand the actions of the gods, Aristotle was moving towards a reliable method of predicting the weather.

Weather's central place in life combined with its intangible nature to give rise to a great deal of folklore and superstition. Much of which is still with us today. Most of it, however, confuses cause and effect. Ancient Egypt's reliance on the height of the divine, life-giving Nile as a sign from the gods of how bountiful the coming harvest would be was not a sign from the gods but of rainfall levels. A red sky a night really can foreshadow a period of calm weather - provided you are at the right latitude. The dust particles trapped by high pressure can deliver a red-ish sky, but the phrase's Biblical origins do not explain that. Seaweed really can absorb moisture in the air, indicating humidity levels, but until humans understood what humidity was, how it came about and what it might lead to, these signs provided scant insight.

Meteorology, for all its importance, did not come anywhere near to being scientifically cracked until the 1600s. As with any science, progress was slow, over decades and centuries, and relied on scientists creating new technologies. It required wider acceptance that weather was not simply an act of God, which itself required a shift in thought really only prevalent with the coming of the Enlightenment. It also required so many complex measurements, and complex instruments to make them, that the science of meteorology had to wait for other sciences to catch up.

Accurate weather measurement, and in time prediction, demanded global records of events big and small over the course of years. Weather measurement and forecasting, even locally, remained largely a matter of conjecture for centuries even to the most rigorous scientific minds. It was only in the 15th, 16th and 17th centuries when hygrometers to measure humidity, thermometers to measure temperature, and barometers to measure air pressure were developed, that changes in weather and climate could be observed. Even then it was hard to communicate observations so what patterns could be observed were largely academic as any warning would arrive too late.

The evolution of human society is frequently a story of technology facilitating advances and vice versa. The winners in the global trade contest would not necessarily be those that invented the technology but those that put it to the best, most profitable use. From the 1830s onwards the telegraph, the first reliable means of communicating very quickly over hundreds of miles, facilitated many advances in commerce, culture, government, war and trade.

By the 19th century few (but by no means all) people of learning still believed weather was solely the whim of a deity. Perhaps appropriately, or necessarily, it was Britain, the island nation increasingly reliant on trade that established the world's first state meteorological forecaster. Britain's weather is notoriously unpredictable (hence the British propensity to talk about it) and in 1854, the imaginatively named Meteorological Office started their work.

Francis Beaufort, the man who would give his name to the wind force scale, was one of the most important figures in the British Royal Navy as the man responsible for their maps and charts. His friend, Robert FitzRoy was given command of a Naval ship, the HMS Beagle. Like Cook before him, the use of a Royal Navy vessel meant fulfilling military goals and he was to map routes around the South Atlantic and into the Pacific. FitzRoy, an amateur scientist himself, was aware that one of the hazards of command was loneliness. He sought intellectual companionship in the shape of a geologist that might provide useful information as well as company on the trip. He approached a few, but none was willing to undertake the long, arduous journey. Beaufort introduced FitzRoy to a young naturalist called Charles Darwin. It was he that would join FitzRoy on his five-year journey.

Two decades after Vice-Admiral Robert FitzRoy returned from the South Seas, he established the Met Office, intended to help sailors and those who made a living at sea. FitzRoy knew that predicting the weather was key to safe passage and survival at sea, as well as military supremacy, and he coined the term 'forecasting the weather'. He took information from 15 land sites around Britain via the new telegraph system, the only possible way to gather information across a wide area quickly. It helped build a picture of storms forming and moving, and enabled information to be passed back and warn others as necessary. As the network of observers increased more readings from further afield created a service to those inland as well as on the coast. Today the Met Office provides weather data to airlines and space agencies, they help track the spread of diseases and plan military manoeuvres. In the mid-1800s they were helping those exploring new terrestrial worlds as well as those planning naval engagements.

Those early forecasts compiled basic data on wind speed and direction, rainfall and cloud formation to build short-term predictions. Today, terabytes of data is gathered every hour via sensors everywhere from space to the bottom of the ocean. Meteorology is an empirical science alongside physics, astronomy and chemistry. It has developed as a science, learning and changing as old methods and theories become discredited. Its predictive abilities, like all of sciences, are acknowledged to be limited or only useful within certain boundaries.

Today, although weather remains often unpredictable on a very long-term scale, global patterns can be mapped and large weather events predicted within reason. How storms and weather fronts form is understood. The accuracy of prediction is increasing as temperature, wind, humidity and such can be measured on an ever more micro scale. It is now very rare that destructive weather events cannot be foreseen at least days in advance and warnings provided.

Those early metrological developments led to efforts to study the conditions leading up to different types of weather. They were examined and recorded, analysed and studied by those seeking to understand more about the world. Often their endeavours were motivated by a simple desire to understand the universal. Sometimes it would for more pragmatic, selfish reasons. Whatever the motive, before people took this approach to weather, and other elements that make up the natural world, science played second fiddle to belief for most people. As far as the public, and often their rulers, were concerned, weather was the earthly embodiment of godly actions. Once that belief was challenged, what would be next?

6
THE AGE OF ENLIGHTENMENT

"A man who is certain he is right is almost sure to be wrong" –

Michael Faraday

The Age of Enlightenment saw science and reason gradually start to replace religion and faith in many areas of understanding. It was ultimately a revolution of science, philosophy, and politics. Whilst religion would continue to hold sway over questions of morality, law and behaviour, its ability to fully explain the world was increasingly questioned, especially by many educated people. In some cases this questioning of God and nature was seen as sacrilegious. Elsewhere it was little more than the eccentric preoccupations of a privileged few. Some Enlightenment thinkers saw their challenging of received ideas of humanity and the natural world as vital in order for society to progress. Others saw it as the God-given duty of humanity to learn and to examine the workings of His incredibly complex creation.

Throughout the 1700s science was increasingly trusted to explain, or at least have a good attempt at explaining, the mysteries of the world. Science was demonstrably improving understanding of the human body and medicine, physical phenomena and chemistry. The scientific method was increasing in its importance and power. Methods of empirical examination, the rigorous testing of theories, repeatable experiments and peer reviews were vital to anything seeking credibility in this new era of reason.

Rationality, empiricism and scientific method delivered results - repeatable and repeated results. In medicine in particular, science saved lives and tangibly limited the spread of disease. This age of reason also affected prediction. Theories and models were tested and tested again, and eventually adopted as ways to predict the physical effects of an action.

As Albert Einstein would observe 200 years later, a theory could only ever be proven wrong. Science embraced this challenge early. Whilst theories could and would be debunked by new evidence, the scientific community simply adapted its work to account for it. If a theory seemed to explain something satisfactorily, it would be adopted until something disproved it. Upon which this new finding would be examined and a new theory presented. Certainty was no longer required for credibility. Adaptation and flexibility were the new, sensible approaches. Learning never stopped.

With this newfound willingness to question and view certainty with caveats, scientific prediction was limited. It always carried the proviso that what could reasonably be expected to happen in a certain set of circumstances would probably, but not certainly happen again. Science, underpinned as it was by logic and mathematics, would seek conditions where a theory could be wrong, or to put a value on the likelihood that something would or would not happen. Science accepts that it could always be wrong. Science, by its very nature, was robust enough to be challenged.

This lack of certainty meant science was not in the business of predicting the outcomes of wars or politics, or of individual lives. Even thought it would affect them it would do so in unpredictable ways. It would contribute to new ways of fighting, new sources of wealth, and great advances in medicine and welfare. It would not offer reassurance or support to a particular agenda. It would undermine many of the predictions of religion and faith. It would also influence the approach of other disciplines, like philosophy, that consider what is and what is not certain.

Aside from weather the other great challenge to efficient, safe ocean-going passage was reliable navigation. Whilst cartographers and explorers had been able to map coastlines with some accuracy, it was hard to judge the speed of a ship, so it was hard to judge distances. Even with a reliable map, precise measurements of speed and distance at sea were almost impossible, and even a small error could become huge when travelling over thousands of miles. It all came down to accurate timekeeping.

Whilst science sought to explain how the world works, those whose concerns was more pragmatic pursued ways to use science to their own particular ends. International trade required safe passage, accurate navigation, and a strong navy to protect its sea-going interests. In the early 1700s there was a series of accidents that saw British naval vessels wrecked on the Scilly Isles off Cornwall. These accidents were put down to navigators being unable to accurately judge their ships' positions.

The British Parliament, along with private individuals, set up a series of prizes to tempt the country's scientists to solve the problem of longitude. The prize to the winner was equivalent to over a £1m today. The prize to the nation and its merchants was a decisive advantage in both trade and war.

Timekeeping has been an obsession for humans certainly since the Agricultural Revolution, possibly before. Those early astrologers and mathematicians measured the degrees by which astral bodies moved day by day. They measured the passing of time by the burning of uniform candles or draining of water or sand through an aperture, or by tracking the movement of the sun. The development of mechanics made instruments that oscillated at a regular rate, an attribute applied to timekeeping. All of these clocks, however, relied on mechanisms moved for a limited period, either by weights slowly pulling a clockwork movement or by winding and then releasing that stored energy through springs.

As important as it was to measure time accurately on land, the country that could accurately measure the passage of time at sea would navigate more accurately, and travel more directly and more safely. Importantly, they would win the global trade contest.

At sea, you could not rely on weights and pendulums. The method that drove the most accurate clocks of the time was rendered useless by the relentless movement of the ship. Wound clockwork was unreliable and imprecise, certainly when it came to using time to measure how far a ship had travelled. British clockmaker and carpenter John Harrison solved the challenge first and best with his marine chronometer, produced in 1735. It took him over 30 years but his pioneering achievement, though expensive to produce, gave British shipping an early advantage, reinforcing its front-running position.

Science had enabled trade, particularly British trade, to improve vastly by analysing, understanding and creating tools in chronology and meteorology. In turn, merchants, and therefore governments, saw value in science and invested in it, seeking commercial developments, new products, and new efficiencies. Trade's international scope also brought new ideas into play, advancing science with new perspectives, new materials not found domestically, and ideas advanced by different cultures.

The trade in ideas was and remains arguably far more profitable, and more significant, than that of sugar, salt or cotton, even oil and gold. Whilst money could be made directly (if shorter-term) from the buying and selling of commodities, ideas could completely and irrevocably alter entire industries and populations. Scientific advances also produced new technologies; technologies made more money, which funded more science and so a virtuous cycle developed.

Importantly science was not just revealing exciting, important and practical new information about how the world worked, it also provided a new way of thinking. It questioned, experimented, theorised, revised, and challenged. It sought evidence and proof, and remarkably it tried to prove itself wrong. Far from being a sign of being impermanent or unreliable, science - true, empirical science - derived its strength from its willingness to adapt when presented with new information. Both flexible and robust, science resists change until new ideas are thoroughly tested, but once they are, science adapts. Science predicts based on its prevailing theories but knows that those predictions are only as valid as the theories, that is, until new evidence emerges.

As it had since the 1500s, science was used by the powerful for their own advantage; a return on their investment, risk-taking and faith in its research, but they would also come to be threatened by it. Although funded by politicians and merchants with political, social and commercial aims, science continued an overall trajectory of benefiting humanity with unbiased, apolitical progress. Those with an ulterior motive might twist science, or ignore inconvenient parts of it, in order to prove their point. However science chipped away at those claiming power over others through divine intervention, and it gave more power to those lower down the class structure.

Religion in particular would use its political and social leverage to both fight its own corner against science and to protect its influence amongst the elite. It would not have modern science come along and undermine its loftier, millennia-old, spiritual authority. But it was not alone in being threated by scientific thinking. If everything people assumed about the most basic workings of nature was now being questioned why not also social constructs?

Just as spiritual leaders valued their position of influence over the powerful, so scientists started to do the same. Just as their priestly forebears denounced ideas that threatened their privileges so science's established elite would come to fight to protect their role as holders of knowledge and influence. They themselves would come under the influence of power and money. It would compromise their independence and see them develop theories that were aligned to certain unscientific priorities. In the long-term, however, the rational won out.

Scientific method, its spirit of enquiry, could be applied to explaining how something moved or changed. Did that mean it could also be applied to other parts of life, including the non-physical? Could it explain or evaluate ideas of morality and social order? Could it control or make sense of this new, complex industrial world it had helped to create? This thinking combined with contact with other societies and cultures to generate new political and philosophical ideas, particularly around human nature and the freedom of the individual.

ENLIGHTENMENT POLITICS

As well as advances in natural sciences, the Enlightenment changed politics in both direct and unexpected ways. God's position as the single, simple explanation to how the world worked had been challenged. So it was that His appointed representatives on Earth, be they priest or monarch, were also challenged.

Everything could now be questioned, examined, and if proved wanting, replaced by something better. There was a flourishing of ideas, particularly around democracy and individual freedom. Works that would influence politics for generations were produced by the likes of Thomas Paine, Thomas Hobbes, John Locke, Edmund Burke, David Hume, and Jean-Jacques Rousseau. With books reproduced internationally these thinkers would change how people saw the world in fundamental ways. They inspired the French and American Revolutions, formed the basis of national constitutions, and even questioned what it meant to be human.

Faith and belief would retain a tenacious grip on the everyday lives of many people. This was especially so when they needed to influence anything outside their direct control. However science and rationalism made in-roads into either replacing or at least challenging many notions of faith and prayer and their position in influencing health, warfare, wealth and social order. Whilst Abrahamic faith accounted for most belief in the world, ideas like astrology also remained popular. When it came to matters of great significance, however, the rational was ever more trusted to deliver an absolute, objective solution.

The culture of questioning power, whether religious, social or political, became self-perpetuating. Figures like Charles Darwin were able to take on even the most sacred and long-held rules of the natural world and humanity's place in it, including humanity's God-given right to dominance over all other creatures.

Darwin's most famous work, *On the Origin of Species*, has become one of the single most important books in the history of human thought. Like all scientific breakthroughs, it was built on the work of his forebears both scientific and philosophical. Science, appropriately, evolves and Darwin's worked owed much to not just biologists or naturalists but also the geologists, palaeontologists, and physicists. Scientists that had also proposed how change had occurred over millions of years creating the world that homo sapiens had inhabited for such a relatively short period.

Evolution and natural selection, like Newton's gravity, was and remains a theory.

A theory will always remain a theory no matter how much evidence and how comprehensively other theories in its field are dismissed. In this evolution is a prime example of scientific method and thought, and what it says about human ideas of certainty.

The fact that evolution is a theory not a fact has seen

it dismissed as guesswork; as half-baked, or as evidence that science does not know as much as it thinks it does. Science is accused of arrogance in its dismissal of other explanations of the world, including the religious. In fact, the opposite is true; in its acceptance of its limitations it shows science is robust and open to changing ideas it has held true for decades, even centuries.

Human knowledge must accept that it is, by nature, limited. Science knows that one-day new evidence may emerge. Just as Newton's ideas of gravity were undermined by relativity and quantum mechanics, so, perhaps, evolution may be seriously questioned. Until then, natural selection robustly and adequately explains life.

Darwin's science, as well as other significant scientific developments of the age, relied heavily on the field of probability. Darwin and other naturalists applied the mathematics of likelihood to demographics, the study of population change and make up. He used these statistical-driven ideas to build up pictures of how, when and where animal and plant populations spread.

Recognised as an area of mathematical concern in the 1500s, by the 19th century French mathematician and physicist Pierre-Simon Laplace had formalised the concept of statistical probability. Half a century earlier English clergyman Thomas Bayes gave his name to a form of probability, Bayesian probability, that described the likelihood of something happening based on knowledge, experience or belief, rather than the actual frequency of that thing happening. It was a human expectation rather than a scientific one. Laplace imposed a strict structure on Bayes more philosophical work and created formulae and models to calculate outcomes given certain historical observations. This was the birth of statistics and it would underpin most future scientific research.

To many living in the tumultuous 1700s and 1800s, scientific advances were merely God-given insights into the God-created world. Similarly, the application of these advances were divine gifts brought to bear by God's greatest creation – humanity. Greater weapons of war, machinery that grew the economy, more accurate predictions of the weather, new farming techniques - these developments were all the fruits of God's work bestowed on His favoured people. They were reward for their hard work, their righteousness; a blessing on their nation and their race. This combined with other divine advantages of geography, geology and climate. Nations saw their success as God-given; that they were chosen to do God's work. That God had, indirectly, given them a better military or the means to make more money. It meant nations felt they were uniquely favoured and that, no matter how brutal or immoral their actions in pursuing their goals, they had God on their side.

This was no more evident that when expanding trade interests overseas merged into expanding national interests, leading to colonialism and empire building. European nations took over whole countries, killing, exploiting, crushing religions and cultures, extracting wealth, and leaving chaos behind. All in the name, ultimately, of trade but under the grotesque excuse that those of the Christian faith had the right to rob and exploit. Why else, after all, would God give the Europeans advantages of such magnitude?

7

FACTORIES, WORKERS AND COMMERCE

Science was changing how people saw, understood and dealt with the world. Trade made the world more complicated and life more fragile. These two factors had seen the West increase in power throughout the world. A further development would marry science and trade together, would make Western dominance unassailable, and would increase the importance of prediction.

From the first century CE, steam's potential as a motive force - something that could do work, that could move objects – had been observed. Limited by material technologies and the lack of necessity to replace human labour, it was not until the mid-16th century that a form of steam turbine was outlined by Ottoman polymath Taqi ad-Din Muhammad ibn Ma'ruf. A century and a half later, working, pressurised steam engines were being developed. These engines, of which Thomas

Newcomen's eponymous engine was the most popular, used steam to create a vacuum that in turn would draw up water. They were frequently applied to the tricky matter of extracting water from mines, the first application to warrant such complexity and expense.

Another hundred years passed, and the son of an aristocratic Scottish family, James Watt, applied his learning, curiosity and privilege to science. Asked to repair a Newcomen engine, Watt instead used his own ideas on latent and wasted heat to create a more efficient, piston driven engine. There followed the challenges of turning theory and small-scale experiment into large-scale working practice. Watt's first manufacturing collaboration went bankrupt, but a new backer with better ironworkers, based in the Birmingham area took over. More exacting tools were required in order to make more robust engine parts with greater tolerances. The development of one thing depended on the development of others, and also on finance. To overcome these challenges a profitable use for the engine had to be found.

Cornwall is rich in minerals. The area has been the site of mines since around 2200BCE. Known as the tin islands by the Ancient Greeks, trade in metals from Cornwall started somewhere between 500 and 325BCE. By the 1700s pewter had moved from decorative and rare and only for the elite, to commonplace. It was used in all manner of everyday items, principally tableware such as drinking steins and flagons, bowls and so on. Pewter is largely made of tin and Cornwall had become the centre of tin production in Europe. Tin had become very valuable both at home and in international trade. Here was the money that might invest in advanced engines.

Cornwall is a thin, rugged peninsular and so is bordered

on both sides by the sea. As higher demands for tin production took their toll, the necessity to pump more and more water, ever more quickly, became a financial necessity. Watt spent years installing his new steam engine-driven pumps in Cornish mines before further improving them.

Watt's steam engine designs of the 1770s and 80s, as well as his other work, are widely seen as a principal force behind the Industrial Revolution. His efficient engines, and importantly, the techniques established for producing and replicating them, would change the very nature of work, conurbations, society, politics, and economics. It would make Birmingham the centre of industry not just in Britain but in Europe. The steam engine would not just replace people as the literal driving force of industry; it would also create completely new products. Machines did not just replicate human work; they developed entirely new methods that humans could never do. Machines could be more accurate and more detailed; they could be stronger and faster than any human. The machines meant new jobs maintaining and operating them. They meant swaths of people moving to cities, working in factories, living in cramped, squalid conditions. It meant new cities, but also the expansion of the existing port cities.

From 1800 to around 1850, England and Wales, the engine room of the Industrial Revolution, saw its population double to 18 million people. The population living in cities rose from 900,000 to 9 million. More cramped living conditions brought health and social problems; diseases spread, as did crime. The ruling classes blamed the working classes for these problems and for a predicted food shortage and financial crisis. At the same time, others decried their exploitation in factories and mines and demanded conditions improve. There were demands for workers to be both protected and represented in power. New political ideas found popular support, including the diminishing of rule by aristocracy towards a form of parliamentary democracy.

The effects of steam technology cannot be underestimated. The machines it brought about, the increases in productivity they realised, changed the world. Previously work was limited to the strength of humans or by the raw natural elements they had been able to harness directly. Industrialisation meant cheaper, better and more consumer goods. It also meant the same for weapons of war. Mines were more efficient with more men dedicated to cutting minerals rather than pumping water or moving minerals. Steam engine technology would revolutionise transport by sea and land, increasing speed and reducing cost, and expanding collaboration and innovation. With Britain at the lead of the Industrial Revolution, the country gained crucial advantages in war and trade.

As mass transportation over sea and land improved and became more common, and as such cheaper, so the international trade in goods changed. Until the 19th century the vast majority of trade, for example in silk or spices, served a wealthy elite. Now bulk commodities could be shipped great distances. It changed how people worked and lived. More people were employed in ports and in related businesses. More people consumed the products brought in. More people were employed in factories processing raw materials for onward export to other markets.

Port cities became rich and important. They became administrative, financial and government centres. Entire economies grew up around ships and their crews being in port for long periods. Nearby factories added to the population and the chaos and the wealth. The cities drew people away from rural life domestically and from around the world; the human fuel for the Industrial Revolution. Land was used for more economically advantageous activities. Food was imported in order to feed growing populations.

In the countries where raw materials such as cotton or sugar were more easily cultivated, life became at best precarious, and at worst horrific. Rulers no longer considered what crops were needed for a population to feed itself but what brought in a profit. Landowners focused on cash crops and farmers became labourers. If workers were paid at all, the scarcity of food could push prices too high for them. Market forces affected the prices of commodities and therefore the likelihood of a labourer to find work. Weather could seriously damage a harvest and the landowner would employ fewer workers. This near-reliance on importing and exporting was in fact the birth of globalisation. It set in motion a centuries-long phenomenon of global movement of goods and people and the resultant cheaper products and outsourced labour.

Britain led the way in combining advances in production and transportation to dominate the most profitable of global businesses; textiles. The cheap mass production of the most popular items, clothes and linens, as well as luxury upholstery, brought untold wealth to Britain and its factory owners. The world demand for cotton products seemed insatiable. Britain pushed more and more resources into harvesting, shipping, processing, weaving, manufacturing and exporting textile goods. The economy boomed, urban populations grew, and Britain used its abundant wealth to buy from abroad what it no longer produced at home.

Through war and trade, both made more efficient by industrialisation, Britain's economic and political rule stretched from Asia to the Americas. At the time, patriots saw this good fortune as bestowed on the British by God - reward for their honest, hard-working philosophy. As such, anything they did for the 'glory' of Britain was also for the glory of God. The actions of the British were blessed. As a people and a nation God had made them stronger, more intelligent, more advanced, more civilised than others. When the Industrial Revolution facilitated international expansion through more efficient ships and a better-armed military, British nationals took land and people in the knowledge that they were superior; that they were on the side of right.

HUMAN COMMODITIES

This sense of superiority was nothing new. Since the 1500s Britain and other European nations were responsible for many brutal, inhuman acts around the world. Representatives of these countries journeyed the world, discovering new lands and new wealth. In order to achieve their goals, in order to bring glory and wealth to their nations (and themselves) they indulged in acts from corruption and double-dealing to the wholesale slaughter of entire populations to slavery and the slave trade. Industrialisation served to make Europe more efficient in these crimes.

No history of global trade, industrialisation or economics should ignore the offense of slavery. The trade in humans stretches back thousands of years, and inexcusably continues to this day. It has been legal for almost as long, only being generally outlawed 200 years or so ago. Wars were fought over a trade that saw human beings reduced to commodities to be bought, sold, and treated as their 'owner' decided was reasonable.

With industrialisation and the ability to mass manufacture fabrics, the textile industry became Britain's principal trade. It relied on a raw material that could not be grown within its borders. Much of the cotton needed would come from America. Raw cotton is difficult to pick, and had to be done by hand. In order to harvest cotton as cheaply as possible millions of slaves were captured and shipped across the Atlantic. Britain felt it was justified in forcing people that they saw as less than human to labour for them; to use them as they would gold or coal in order to further their own wealth.

Having made millions, billions in today's terms, not just from the products of slavery, but also from the actual trade in slaves, Britain withdrew when it became clear there were cheaper alternatives. The machines that were fed by planation slaves would eventually replace and ultimately free them. As machinery started to become mobile, it would soon be cheaper to employ trained machine operators to harvest crops than it would be to deal with humans working against their will.

The Atlantic slave trade was at his height from around 1750 to 1850. It was renowned for its horrors, from the capture of slaves to their inhuman transportation in conditions of filth, disease and death, to their treatment at the hands of plantation owners across the Americas. In the course of the hundred years of the trade over 12 million slaves were shipped across the Atlantic, primarily from Africa. Roughly one in ten slaves would die before reaching land (although no one is sure due to of the lack of proper records). With humans reduced to the level of commodities, the price of a slave was directly affected by the price of that which they would eventually harvest, be it sugar, cotton, tobacco or a number of other cash crops.

The American Civil War was fought primarily, although not exclusively, over slavery. Not just the morality of slave ownership but its economic implications. The economy of the American South, a hugely successful and growing economy, had become almost entirely dependent on slave labour. Slavery ensured high profits from one of its principal crops, cotton. That labour would be the economic engine that would drive the Southern states' push into the newly opened west of America. The North, fearing the South's potential dominance in this new, potentially rich land, and being less reliant on agricultural economics, struck a pre-emptive blow. A bloody, brutal four-year war resulted in a victory for the North and a political and economic dominance that remains to this day. That victory also, in time, hastened the end of slavery.

In the early 1860s, in the midst of the Civil War, Southern cotton exports were blocked from travelling through Northern ports to Europe. Manufacturing destinations, in particular the north-west English county of Lancashire, were embargoed. This resulted in the Lancashire Cotton Famine which saw huge parts of industrial England shift rapidly from some of the most affluent in the country, to some of the poorest.

Lancashire cotton makers appealed to Egypt to increase production of their cotton to try to compensate. The Egyptian government refused, suggesting there was no need because vastly increased cotton prices, a simple question of supply and demand, would increase production anyway. It did, production rose by 400% and cotton became Egypt's main export, improving national finances hugely. The trouble was that in order to cope with the demand, the farmers employed slaves. African slaves. Treated just as poorly as those shipped to another continent.

European dominance of global commerce was built on slavery and reinforced by the Industrial Revolution. Ingenuity and science enabled the revolution but the money to accelerate it would come from cheap commodity prices on the back of slaves. The capture and enslavement of humans from one continent supplied the foundation for the dominance of another.

Religious morality and Enlightenment ideas about humanity, equality and the freedom of the individual provided coherent, moral arguments and an intellectual foundation to end slavery. Campaigners at all levels of society across Europe demanded the end of the practice. For all the moral pride some in Britain held, and continue to hold in its part in outlawing slavery, it is worth remembering that economics was a dominant factor in the decision. It is true that many wealthy parliamentarians gained through the slave trade and the more brutal excesses of international trade. It is also true that in outlawing it many lost money. However they were handsomely compensated by the state, and they would go on to earn even more through new forms of commerce.

Advances in technology, the economic impact of American independence, as well as Enlightenment thinking, all played a significant role in the ending of the slave trade. Yet slavery continues today. People are trafficked not to pick cotton, but exploited by criminal networks to staff factories and farms, to produce illegal drugs, and work in the sex trade. It may take further developments in technology, economics and in thinking to end the trade forever.

THE RISE OF THE CORPORATE ECONOMY

The social and economic world order was shaken by the Industrial Revolution. The flow of raw materials and manufactured goods, and the processes in between, meant much more trade being done in many more items. The search for new markets and new, cheaper sources of commodities saw ever more exploration in the name of the private enterprise, the nation state and their rulers. With the increased flow of goods came an increased flow of money.

Whilst much of this wealth flowed to those who already had some, there was a knock-on effect. The standard of living of even many of the poorest of Western Europe was lifted. They had greater access to the necessities of life. Their housing and conditions were improved, even if it was only marginally in some cases. Education became more widespread as poorer people saw the long-term benefit of paying to educate their children. With the rise of factories came more skilled and clerical jobs, which required workers with some degree of education.

With education people became more independent in their decision-making and more influential amongst their immediate peers. The general population could read posters, pamphlets, petitions and even newspapers. They had access to information they simply did not have before, and even those without education heard speeches and lectures from those who did. Gradually populations could no longer be dictated to in quite the way they once were. They no longer just accepted what they were told by the political, commercial or religious elites. Labour was partially replaced by intellect and automation. Production added ever more value to a raw material. Most people worked for an employer rather than for themselves.

By time mass industrialisation had started to take hold of the economy in the late 1700s and early 1800s, the East India Company was 200 years old. It was firmly established as Britain's pre-eminent corporate institution with fingers in many commercial pies and important connections to politics and royalty. Its most profitable cargos ran from spices to dyes to saltpetre (a key element of gunpowder and therefore vital to any military power, including the EIC's own army). Naturally it also included cotton, which, along with dyes, gave it a particular interest in the production of textiles.

The EIC initially harvested cotton in India and produced fabrics and cloths in their own factories in the country. Lobbying by factory owners in the north of England put heavy restrictions on the importing of manufactured cotton goods. So the EIC turned to importing raw cotton, bringing in thousands of bales. The sudden increase in the supply of cotton necessitated new machines that could process and weave it much faster. With these new innovations came a dominant new industry in the north of England. It saw the cities of Manchester and Liverpool become some of the wealthiest in the world.

Factories and automation created a cycle of prosperity; the more money being made, the more was invested in innovation to improve weaving speed and quality, and reduce waste and manpower. With more money came more disposable income, therefore greater demand for imported goods and materials, which generated more money and so on. Underpinning all of it was trade. European economies were increasing dominated by trade and as such the welfare of populations became dependant on the vagaries of shipping and commodity prices. At the heart of this was specialisation – focusing efforts and investment on just a few industries, becoming very good at them, and allowing trade to supply needs outside those industries. Cotton manufacturing became Britain's main specialisation.

European economies were now in the control of businesses; principally trading companies and the private factories that they supplied and that supplied them, as well as the merchants and banks that funded these exchanges. In Britain, it only furthered the East India's power. These were pioneering times; much of the work the trading companies were undertaking had no rules because no one ever thought they would be needed.

The East India Company's unprecedented work meant it usually wrote its own rules as it went. The British government, the only potential restraining force on the Company, either turned a blind eye to its actions or were baffled by its complex structure of shipping, finance, investment and its exploitation of legal loopholes. Whilst some of the EIC's actions were morally questionable, even by the standards of the time, some were seen as outright illegal, dishonest or corrupt. It brutally suppressed local populations around the world with its private army. It colonised territory and forced local leaders to surrender to its rule.

When the Chinese proscribed the trading of opium in the country, the British and the EIC circumvented the law, smuggling almost 1,500 tonnes of the narcotic via India. The slave trade, arguably one of the worst stains on European morality in all history, remained a profitable business for the Company, even after the trade itself was outlawed in Britain.

The East India, and others, did much of this with at least the tacit knowledge of Parliament, largely because many in Parliament were shareholders. Others relied on the factories and their owners for political and financial support. Where once an aristocratic and religious elite ran most European nations, now it was joined or even superseded by an interlaced network of commercial wealth and political power. Factory owners, merchant and trading companies, banks, parliamentarians, the military and lawyers collaborated for their own ends.

The dependency of the national economy on a few private businesses holding a virtual monopoly on manufacturing and trade put governments in a weakened position. The private, corporate interest was now seen as synonymous with the national interest. Without the tax and employment these companies provided, there would be social and political collapse. It would become necessary to understand more and more what they were doing and what might become of them.

Parliament drew up acts specifically for the benefit of private companies. In one instance it was forced to bail out the EIC, so dependant was the British economy on the Company, and so powerful and important were its friends.

Holding and administering the many territories of the East India was expensive and frequently precarious. In Company ruled Bengal, a famine (later known as the Bengal Famine of 1770) killed an estimated ten million locals. It damaged the EIC hugely, both financially and reputationally. A combination of disease and drought, compounded by the Company's increasing taxes and demands for cash crops, decimated Bengal's population. Many of those that survived fled in search of food and away from the Company's grasp. The East India was struggling and needed financial help. It appealed to the British government for assistance and a compromise was reached. Rather than direct financial support, tariffs on the Company's tea imports to America would be lowered. In doing so they undercut local American traders, which precipitated the Boston Tea Party and as such fired the starting gun to a revolution that saw America become an independent sovereign nation.

From thereon in the EIC came under increasingly strict regulation and scrutiny. In 1773, the year of the Boston Tea Party, Parliament passed an act that saw the government take greater control of the Company's assets, including the country of India. However it continued trading for another century and would not expire until the Indian Mutiny or Rebellion. Local resentment of Company rule, by then over 250 years old, combined with disquiet amongst the hundreds of thousands of Company soldiers recruited from Indian territories. Hundreds died on both Indian and British sides, and both were guilty of atrocities, but it saw the East India's rule finally ended, albeit replaced by British imperial rule. What was left of the EIC's assets and holdings were either sold off, or in the case of territories, armies, staging posts and routes, nationalised and subsumed into the growing British Empire.

What is true of the British Empire and its collaborators is true of many other European nations, especially the Portuguese, Spanish, and French. Trading companies played similar roles across South America, Africa and East Asia. They acted as proxies or scouts for sovereign empires. The long, bloody history of European colonisation is well documented. As are the economic and political reasons behind them, whether it was exploiting resources or protecting trade routes.

The wealth of both European nations and of powerful individuals within them shaped the world for generations. A long and shameful history of bloodshed and exploitation formed the basis of the economic successes of a few fortunate countries.

FACTORIES, WORKERS AND COMMERCE

A fairly insignificant island off Western Europe continues to be one of the wealthiest nations on the planet despite producing little because of the events of the 1700s and 1800s. It did so through industrialisation. Britain moved from being self-sufficient to being a manufacturing, and therefore a trading nation. This complex, global marketplace of nations specialising in certain products and industries needed analysing, explaining and predicting.

8

THE STUDY OF WORK

By the 18th century trade and manufacturing provided the foundation of most European economies. Success in these areas, therefore, became the principal concern of political rulers around the world. The increasing number of businesses, products, commodities, workers, goods and consumers also brought complexity. Government as well as banks and administrators needed to understand the movements, causes and effects of money and commerce. What policies, what investments should be considered to affect positive change and growth? What could they do to increase production? How to measure economic success or failure and predict the effects of movements in prices and labour?

The study of what might be called economics (the study of supply and demand and markets) has its earliest recorded origins in Ancient Greece. Around 700BCE the poet Hesiod wrote the poem *Works and Days*. He was the first writer known to reflect on the relationship between work, production and scarcity in agriculture. The poem also covers how to farm effectively, the moral value of hard work, and the injustice of a ruling elite that takes more than it provides. A century later his compatriot Xenophon wrote the *Oeconomicus*, coining the term economics and defining it as the effective and successful management of the household and farm. It also expands into social affairs, comparing rural and urban life, and the place of religion and education both in the home and in the wider state.

Around 550BCE, in the Chinese kingdom of Yue, Fan Li made even greater strides towards establishing economics as a discipline. Born poor but the descendent of emperors, he went on to become one of the most influential thinkers of his era. His career in politics stretched from introducing ideas of efficient bureaucracy to using a form of psychological warfare. In the latter, he persuaded the most beautiful woman in Yue, Xi Shi, to ingratiate herself with an enemy king, known for his womanising. In time she manipulated the king into isolating himself. He ruined his state, killed himself and left his kingdom to the mercy of the Yue kingdom.

With this success Fan Li left public service and settled down (with Xi Shi) to a life of commerce. His medicine business was hugely successful, and allowed Fan Li to write extensively on science, strategy and business. He is known as the first writer in history to commit to print specific rules for success in business, although many of his lessons were arguably based on his political experiences. Some of his teachings are still espoused today, from realising the potential of employees to understanding the nature of your customer to being wary of price wars. Some now seem obvious, like making sure you collect your debts and being astute when buying stock. Undoubtedly, however, this was one of the first times anyone had written on the nature of commerce, supply and demand.

From these early studies grew ever-broader treatises on how important money, work and commerce was to the wellbeing of the state. With the emphasis on agriculture as the source of both personal and national wellbeing, early work was focused more on successful land husbandry. Later thinking related this success to success in trade and war, and therefore in politics.

In Plato's *Republic*, his ten-volume work of around 370BCE, the philosopher reflected on the nature of humanity and of an effective nation state, as well as the interplay of philosophy, labour and money. His fellow titan of Western Philosophy Aristotle also penned an equally expansive multi-volume work *Politics* a decade or so later. In that, he defined forms of rule and government and in doing so started to outline early economic models. *Politics* examined private property and common ownership, democracy and the morality of greed.

Around the same time in India, the *Arthashastra*, a multi-authored Sanskrit work (although principally attributed to the philosopher and royal counsel Kautilya) was taking shape. The title, translated roughly as 'the science of politics', refers to the effective running of a country through law, government and royalty, popular consent, war, tax and economics. As in Greece, in these simpler times economics was almost entirely centred on good farming practices with some reference to trade, law and social order. The book defined the moral duty (or dharma – the virtues which enable life and growth) of rulers to protect the population and to act justly. That they are responsible for, amongst other things, ensuring the people can produce enough to survive. That people will be compensated for losses due to war or theft, and will be taxed only according to their means. It also covered the state's duty in exploiting mineral wealth, establishing mines, formalising weights and measures, and creating standards. With all of this it laid the foundation for a nationwide, uniform and trustworthy market.

As exploration beyond national borders increased, so did the spread of religious and political ideas, as well as trade and conflict. Economics became a key, and ever more complicated part of good governance of any empire, state or region. It provided the link between ruler and the ruled and the welfare of both. It was the difference between poverty and comfort, peace and revolt, failure and success. A healthy economy even provided some barrier to, or at least compensation for, the impact of natural disasters. The gods still held ultimate sway; they could still destroy a nation without warning, but a wealthy, well-governed nation would be more likely to weather the literal or figurative storm. A well-funded state was well prepared, and it seemed, favoured by the gods. Increasingly wealth was married to moral good, at least in the minds of rulers. Fortune, and as such God, had favoured the rich, although the Abrahamic religions would seriously question this assumption.

The subject of economics, although not referred to as such, occupied itself with questions of government, money and work. It analysed the most effective, and sometimes most moral choices. It challenged and proposed alternative methods of government or different ways to organise society to achieve certain aims. It examined the nature of cost in times of plenty and of scarcity. It pondered upon how one could put a value on work of differing types. As trade grew and agriculture, though still vital, waned in the list of rulers' priorities compared to, say, the possession of mineral wealth, ideas of collective purchasing and markets came to the fore. Throughout all this, however, economics was still very much the concern of the philosopher, the thinker and the sage. It was for rulers to choose to accept or dismiss their worldview, their opinions of human and economic behaviour as being the most convincing counsel. Some rulers would see these ideas as no more than intellectual theory; a high-minded diversion of little practical application. Others found precious guidance in their attempts to steer their complex, multifarious state towards security and prosperity.

For much of human history the broad ideas around wealth and productivity were fairly simplistic and philosophical. In 14th century Tunisia, historian Ibn Khaldun would become the next great influence in economic matters. Another member of the upper echelons of society, Ibn Khaldun could claim to be a descendant of one of the Islamic prophet Mohammed's companions. As such, he received the best education and he entered politics, where, due to the febrile nature of the times, he explored foreign lands, was imprisoned, climbed the power ladder, and fell back down again

Documenting his own dramatic life as well as the history of humanity to that point, he recorded the lessons he had learned about the world, politics and people. He codified ideas such as social cohesion and how a social group, if studied, can reveal the keys to both its success and its future downfall. This, he stated, would be true of a small tribe or a mighty empire. In looking at these complex ideas, as well as themes such as cultural assimilation and appropriation, Ibn Khaldun effectively founded the discipline of sociology - the study of the patterns and interactions of a social group. In doing so he also formed the basis of how outsiders could predict how and when a social group might succeed, decline and what might replace it.

Another groundbreaking idea was that the economy was dependent on the addition of value. The more work and skill that was added to a basic commodity in order to create something, the more its monetary worth. Additionally he noted the contrast between sustenance – what you need to survive and continue – and profit – what was left beyond that. Ibn Khaldun also suggested that money must have an intrinsic, rather than symbolic value, and as such should always contain metals such as gold or silver.

Although still not a singular discipline of its own, the study of economics would increasingly concern itself solely with money – its nature, use, production, movement, value and development. As a result, the idea of capital started to immerge. Capital is in many ways the opposite, but also paradoxically the progenitor of money. Capital was anything that facilitated the creation of more money or any other addition of economic value. It could enable someone to work (a tool or workshop) or work more efficiently (a horse to draw a plough rather than human effort). Capital might be something that will increase in value due to the forces of supply and demand. In short, it is almost any physical asset other than money. Money, by contrast, was static, never growing or increasing. Economic theories changed significantly with the rise and ultimate dominance of a concept called mercantilism.

The concerns of rulers were traditionally limited by geography, only ever really concerned with their, probably quite limited, nation. By the 16th century, particularly in Europe, global trade demanded rulers possess a bigger perspective. Europeans had occupied the Americas and established bases of power in Asia. National rulers sought international status. This meant the deployment of national military forces, sworn to protect the population and the monarch, now charged with expanding the national economic interest. By this stage what was good for the economy was good for the nation and good for the monarch; it was noble, godly and honest. It was a duty and a privilege to die in service of one's nation.

Mercantilism was a political and economic theory that proposed the nation's economy, and therefore the nation itself, was best served by the ongoing accumulation of treasure; specifically, gold and silver. Adherents of mercantilism held that, as long as trade was central to economic wellbeing, international trade in goods and products was limited in its effectiveness. In modern terms, the view was that trade was a zero-sum game - if one country gains, another must lose. Treasure was the only absolute form of wealth and it should be the nation's priority to gather as much as possible for itself, thereby depriving others.

This view meant states enacting a policy that discouraged the importation of goods (and thereby giving money to other countries) and encouraged exporting (thereby taking their money). This led to, along with other policies, the imposition of trade tariffs on those importing goods into a nation – effectively charging them for access to a market. Those tariffs further reflected the priorities of mercantilism that preferred the import of raw materials, which would be worked on and then exported at a profit. The policy also put pressure on the importation of anything the nation produced in sufficient quantity to fulfil its own needs.

Mercantilism's emphasis on the national accumulation of wealth had an international influence - the wealthier a nation overall, the more outsiders would want access to the market, which meant them paying tariffs, which added to the nation's wealth – a virtuous cycle. It also demanded that a nation should maximise its economic output using every available resource (land, people, buildings) to produce as much as possible either for domestic consumption (to reduce the need for imports) or for export.

The pursuit of treasure in the form of import tariffs, as well as other types of taxation, brought a burgeoning national bureaucracy, which had to be paid for. It also demanded a strong military structure to enforce tariff payment, or to blockade untaxed goods. It therefore meant a balance between how much tax could be raised versus how much would be spent on administering and enforcing them. This resulted in increasing complexity in terms of planning, accounting and budgeting on a national, and later imperial level. Despite this new level of complexity mercantilism represented a revolution in economic thinking and it was fervently, and forcefully, adopted by the trading nations of Europe including Holland, Britain, France and Italy. It was not without its opponents, but it would dominate economic affairs in these countries for almost 200 years and its effects would continue well beyond that.

The concept of inflation also appeared in the 16th century. French political philosopher Jean Bodin wrote extensively on history, particularly the place of politics in historical developments. He also wrote about the technical aspects of history; the importance of hard facts and the analysis of data as a key to understanding the nature of a society. His contribution to economics was his realisation that money supply changed due to the importation of gold and silver from overseas, and with it, the prices charged for goods and services changed. He drew in other factors such as population growth, waste, trade and migration, and he ultimately outlined the basis for the principle that prices reflect the amount of money in circulation (known later as the quantity theory of money). The huge importance of this thinking was the implication that treasure hoarded actually decreased in value as it rested in a vault unchanging whilst prices increased.

As more competing economic theories emerged, they also drew in alternative views of morality and political ideology. The role of wealth as fundamental to the wellbeing of a nation and its people led to new ideas of how that wealth should be used and generated, what fairness was, what money was worth and what productivity meant. New ideas emerged concerning tariff-free trade, consumer demand, regulation and how growth could (or even if it should) be managed. Economics, although still not a common term at this stage, still concerned itself with a broader, theoretical analysis of the socio-political world.

THE PHILOSOPHY OF ECONOMICS

In his best-known work, *Leviathan*, the English philosopher Thomas Hobbes, examined the relationship between state and population; between ruler and ruled. Hobbes lived through a turbulent era that would see the execution of a king (Charles I), a civil war, the establishment of a republic in Britain (under Oliver Cromwell), and the subsequent re-establishment of the monarchy (in Charles II). Hobbes used these history-defining events to reflect on human nature and what was required to govern and rule over it effectively. He argued that humanity was inherently, unalterably selfish, motivated by personal gain; not a social creature inspired by community and altruism. Rulers and their laws must account for this selfishness and seek to prevent people from doing harm, to themselves and to others; to keep them peaceful and comfortable. Humanity tends to seek power - in which Hobbes includes wealth, status, knowledge and honour - but people must relinquish or never hold some power, transferring it to those with the right skills – to rulers. Those rulers are then charged with the duty to protect the people from themselves.

In the late 1600s Hobbes and his near contemporary John Locke established that humanity was part of the natural order of things; not superior to it, ruler of it, or in any way separate from it. As such humanity should be studied as nature would be studied - that is empirically, and over time, objectively. Hobbes and Locke also introduced the notion that humans should be free - encouraged to think for themselves about their place in the world and what it is to be human.

Sir William Petty, an English scientist and one of the first people to later be described as an economist, brought together two key influences in his life – the work of Hobbes and the scientific, empirical methodology of Francis Bacon. Petty served in Oliver Cromwell's army in Ireland as a medic but moved to 'new sciences'. A respected and self-made man, he became a landowner and MP. A measure of his ability and knowledge he retained an influential role under the restoration rule of Charles II. Whilst in Ireland, Petty developed methods of surveying the land and the people and thereby understanding what these two could produce and what tax could reasonably be levied. A pioneer in economics, by the end of the 17th century he had established theories on money and monetary supply, interest rates, and investment. He also advocated minimal-to-no interference with the market by government.

Naples is one of the oldest trading cities in Europe. The Greeks settled there around 2000BCE and from thereon in it became a hub for shipping and commerce. It became the most powerful city on the Mediterranean and retained that position well into the 18th century. The city provided the perfect laboratory in which Ferdinando Galiani could study trade. Like many educated men before and after him, a career in the church provided Galiani's initial education. A noted wit and intellectual, before he was 25 he'd produced two widely-lauded works, one of which, *Della Moneta*, would be recognised as the first study of economics ever written.

In 1751 the *Della Moneta* set out what money was and how it came to be. Galiani asserted that money was the natural result of the necessity of trade. In doing so he was, by almost a century, the first thinker to formalise the study of economics. He examined the importance of money to governments; how the value of any product or service directly related to how useful it was to people, and the economic effects of supply and demand.

Galiani also went on to produce another influential economic text, *Dialogues sur le commerce des bleds*, or 'Dialogues on the commerce in wheat'. In 1769 he set out his theories on free-trade. Being convinced that there was no uniform system of trade which could work effectively for all nations, Galiani also believed that trade was a zero-sum game – if one nation gains, others must lose the same amount. Despite a drive towards scientific rigour in his analysis, Galiani's economic works were essentially philosophical in nature, using the thought experiments, dialogues and fallacies more common to that discipline.

Isaac Newton, chief amongst his contemporaries in the Scientific Revolution that foreshadowed the Enlightenment, dedicated his life to dissecting the natural world. Through intellectual brute force, original thinking and rigorous method he examined the everyday elements of light, sound, heat and motion to understand how they worked and predict what would happen to objects under their influence. Newton was a mathematician and his work proved that the mechanisms of the natural world were underpinned by maths. So much that had previously been put down as the inexplicable way of the world, the products of divine creation too mysterious for humans to even hope to understand, was now defined by means of an equation.

Just as the Enlightenment radically changed views on science and what it could achieve, it also changed philosophy. How the natural, physical world worked may have been the concern of science, but how everything else worked, and whether it could work better, was the concern of thinkers. That the nature of God, humanity and the fundamental ways of the world should be questioned at all was a philosophical matter. Questioning authority and the long-accepted 'natural' order was at the base of much Enlightenment thought. Just as science constantly questioned itself, so too should morality, politics and human experience.

The Enlightenment saw the rise of many brilliant, iconoclastic thinkers. One of the chief concerns of many Enlightenment thinkers was the nature and limits of freedom. The freedom of the individual to think and act as they choose, a person's relationship to authority, and the rights of that authority to rule and limit freedom.

Some thinkers, like French writer Voltaire, published stories, allegories, even satires on how people thought, their attachment to money or social standing, their fear of the unknown and ready acceptance of conventional views of morality. Others including David Hume wrote more conventional philosophical treatises, notably proposing that the empirical methods of science might be applied to human nature, morality and the mind.

German philosopher Immanuel Kant took a similar approach, but with a more pragmatic view of human experience and thought, as well as the rights a person innately has and how best to use them. Whilst Thomas Paine and others penned more inflammatory, direct challenges to the social and political status quo. This age of tumult in the world of thought, politics, science and literature had many tangible effects, not least inspiring the French and American Revolutions.

By the late 18th century, political economics was establishing itself as a credible discipline dedicated to understanding the relationships between trade, production, government, society and law. It considered what made a state wealthy or impoverished, and the effects that had on its rulers and its people. These ever more complex relationships needed analysis. No longer could a ruler just declare their will, their judgement over simple, binary matters. Complex questions demanded specialist understanding. By analysing these matters they could be streamlined, adjusted and importantly, predicted. By this stage, ruling a state was less a matter of worrying about divine intervention and good fortune and more about social planning and the accumulation of wealth.

Every so often a work or idea goes on to have repercussions well beyond its original design or expectation. Something emerges that is both of its time and well beyond its time. Adam Smith, a Scottish philosopher, started thinking and writing about society, morality and freedom around the time of the Industrial Revolution. As the realities and the possibilities of industrialisation started to become clear across Europe, Smith wrote the seminal *The Wealth of Nations* (fully titled *An Inquiry into the Nature and Causes of the Wealth of Nations*). Published in 1776, Smith's book analysed how a nation establishes and expands its wealth. He introduced important new concepts around labour and productivity, asserting that an economic system appears of its own accord, naturally and without need for outside interference. Like nature it finds its own way to be efficient, but is constrained by barriers of taxation and self-serving entities. Smith also predicted (but did not strictly condone) the way factories would divide up labour amongst its workforce and how innovation would be key to improving efficiency and profit.

The Wealth of Nations made Smith's work a foundation for almost every economic theory that came after him. Seen as the father of modern political economy, *The Wealth of Nations* was the first formal study of national wealth and trade as it related to law, government and social convention. It proposed, and convinced many, that the only sensible, practical, natural basis for an economy was to view both individuals and nations as inherently, intransigently self-interested. That self-interest ultimately drove all economic activity. Bearing that in mind, however, like Hobbes, Smith's aim was to mitigate that instinct and improve the lot of the mass population; to improve "the human condition in practical ways for real people". He reflected on the wider political and physical effects of trade, and on morality and ideas.

Whilst it is often overlooked, Smith did make an important distinction between wealth and profit. The enrichment of the nation, society and the individual is vital, but that cannot be done through the singular pursuit of wealth. In Smith's view what mattered was that profit was returned to the economic system as capital. Only by reinvesting profit by employing more people, training them to be more efficient or to produce goods of a higher value, or by buying bigger and better facilities, could an economy thrive. It challenged the moral monopoly held by the Abrahamic faiths that had always condemned the pursuit of wealth as self-serving, corrupting and immoral. Wealth, in Smith's view, is not the goal; capital is. Capital is morally superior because it helps others to help themselves.

Smith also proposed that by accessing a greater market or greater number of markets, a country could dedicate more labour to their particular specialisations. That in turn means they could make more of their specialised products, thereby earning more by selling them, and using that income to improve their skills of specialisation even further. In short, trade enables a person, group or country to 'realise their potential'. It is an idea, like many of Smith's ideas, which would have significant implications a couple of hundred years later.

The implication of much of this was that labour - the work of men and women - is the underlying, root source of a nation's wealth. The more efficient that labour, the more people available to work, the greater the nation's earning potential. A simple enough relationship complicated, however, by the interdependence of wages, profit and rent. An increase in one of these without compensating in others would result in potentially serious problems of society and productivity.

Mercantilism, the goal of increasing national holdings of currency wealth, was the favoured economic practice of Smith's era, and he was a prominent opponent of it. He saw the duty of national government as maximising productivity and freeing its people; allowing them to realise their own economic worth. If a nation can maximise its labour, and thereby the amount of what it produces, it must therefore maximise its wealth by maximising its access to markets at home and abroad. According to Smith, regulation and restriction in any form, including taxes and tariffs, is antithetical to economic growth, and as such the wellbeing of all. A civilised society is a trading society, crucially a competitive trading society based on need and self-interest. This and only this would realise the most amount of good for the greatest number of people.

The established order, those with a vested interest in the status quo, however, sought to continue to boost exports, limit imports, and increase the national hoard of silver and gold. Smith believed that real, robust, long-term wealth came from economic transactions, the movement of goods and services and currency. This complex set of interactions, rather than hoarding, would eventually become the dominant measure of a nation's economic success – its Gross Domestic Product or GDP. A simple number that would measure all of the sum total of a state's economic efforts and the effective stewardship of the government.

This also marked a sea-change in who ruled a country and what their economic priorities were. Power was inextricably shifting from the aristocratic and monarchical elite towards the moneymen. Merchants, bankers and factory-owners were both the measure and drivers of national prosperity. Rather than being appointed by God, they had achieved their power through work and risk. They would not just represent a new ruling elite, but a new way of viewing power.

Economic movement – the flow of goods, money and labour – was fundamental to Smith's view of national wealth. Long-term prosperity meant the movement had to continue, and continually increase (otherwise it would be overtaken by inflation). This flow of transactions would gain its own momentum and economic growth would inevitably occur, provided that movement was unimpeded.

Key to this continued, self-perpetuating growth was the ongoing growth in productivity - the amount of work put in against the amount produced. This was particularly relevant in the industrial age where such factors were no longer limited by the physical capacity of a human worker. That said, arguably Smith's most famous contribution to economic thought is the idea of the division of labour. In order to illustrate his theory that allowing individuals to specialise in a single task, rather than expect them to take on a comprehensive role in the making of a product, he considered a pin factory.

A menial but necessary task of the time, Smith suggested that an individual would be hard pressed to make a single pin in a day. By dividing each element of the pin-making process into separate tasks and assigning them to separate, specialising individuals, ten people would produce 48,000 pins a day. Not only would this make the factory more efficient, but it would enable all of those involved in the production to understand their role not just within the business, but in increasing wealth within broader society.

With the increase in production such practices would introduce would come a surplus of goods; a surplus which could be traded. Increasing productivity would lead to a virtuous cycle of greater profits, greater wages and greater consumption, which feeds greater demand, productivity and so on. Furthermore, the increase in profit would see that profit turned into capital - greater investment in more efficient production methods and equipment, which in turn would create more wealth. This would only be the case if the owners were secure in the knowledge that that capital was their own, safe from state interference.

A national government's sole responsibility, for Smith and his followers, boiled down to law and order, security, infrastructure and education (all of which had a vital role to play in the creation or protection of the wealth of a nation). Government would leave the market and those that supplied it open and free. Limiting government's role even just to those areas of law and wellbeing still required funding, which inevitably had to come via taxation. That aside, government could, and should, sit back and watch the economy feed itself through consumption, productivity, innovation, profit and reinvestment.

This cycle of growth was effectively automatic and required no further input – the market would boost the production in desirable goods and eliminate or reduce those that were not wanted. Where there is scarcity in the market there are higher prices, as such higher profits, so more capital is accumulated and invested in better production which leads to higher profits and so on. Where there is plenty or surplus, prices and therefore profits reduce, producers move out of that market and invest in areas of scarcity instead. This, according to Smith, is self-regulating and does not need state interference.

Free-trade and free-market competition is essential to this cyclic growth. Protected monopolies, regulations, subsidies and taxes skew the system; prices are artificially or unnecessarily inflated, and the poor, in particular, suffer by not being able to afford these goods. Consumption falls and that affects supply and demand, scarcity and profit.

In many ways these phenomena were seen as physical

barriers; frictions impeding the free movement of goods and money. Tempting as it is to draw parallels with the physical world, this type of movement is very different to the physics of motion; the thermodynamic forces involved in the natural world as explored by Newton. Money, value, trade, profit and government are human constructs. Humans wrote the rules, either specifically or through a gradual, shared understanding. Physics is fundamental, irrefutable, and unchangeable. Science necessitated constant questioning and empirical evidence. Science demanded that Newton's laws and observations were questioned and examined. Over the decades an understanding of atoms and later quantum theory changed how Newton's work was seen and even when and where it held true. There were times when the models and formulae would reliably predict an outcome, and times when it clearly could not. The same could not be said of economics and economic activity.

9

HOMO ECONOMICUS

Before the Agricultural Revolution, humans survived by spotting and seizing opportunities – for food, for shelter, or to avoid danger. Agriculture brought a little more safety and certainty, but it meant more work. Food and shelter was no longer a question of nature providing and humans stumbling upon it; they had to be built and cultivated. More toil came to mean greater reward. The other side to this equation was an increased reliance on factors outside of human control, such as the weather. Growing crops and building homes was one thing, but how could humans try to influence the mysteries of nature? God would fill that role. God would control that which was beyond human comprehension. Earthly success for humans therefore became a matter of hard work and righteous behaviour.

If success for the individual was a combination of physical labour and seeking divine favour, so it was true for the collective; for the settlement, and later the nation. It became clear over time, however, that some worked very hard, behaved according to the rules of their religion, but still did not experience individual success. They and their work remained important to the collective, to the nation, so their lack of comforts or wealth had to be explained.

Religious leaders and their political partners agreed that if earthly success was not forthcoming because of circumstances outside of a person's control; that was God's test. God and country or tribe became synonymous. Demonstrate fortitude, courage and generosity towards your God and your country, the individual was told, and your rewards would come eventually, in this life or the next. Work was moral, improving, character-building and necessary. To dedicate or even sacrifice one's life for one's country was noble and would be rewarded in the collective memory of those left behind and the spiritual world of the afterlife. In truth, the ever-growing nation state simply needed people to work and keep working in order to function.

Adam Smith's *The Wealth of Nations* went on to form the cornerstone of Western capitalism. Published, as it was, at the start of the American War of Independence, this new, vibrant thinking unsurprisingly entered a philosophical mix that would be the basis for a national, written constitution. Founding Fathers Benjamin Franklin and Thomas Jefferson were heavily influenced by, as well as influential on, the European Enlightenment. When the time came for them to establish a new nation virtually from scratch, the ideas coming from the continent of their ancestors were resonant and timely.

Perhaps more than any other single work (with the exception of Paine's *Rights of Man*) *The Wealth of Nations* informed the prevailing American philosophy of post-independence government. The Declaration of Independence's famous assertion of the unalienable, irrefutable right of everyone to "Life, Liberty and the pursuit of Happiness" would, for many, imply the freedom from government interference to achieve financial happiness and security. Of course, being written at a time when not every eventually could be accounted for, what this right really was in practice would be widely debated in years to come. Especially when it came to questions of equality in race, gender and socio-economic background.

Irrespective of its originators' intentions, the Declaration has been used to assert an American's freedom to make money; to maximise one's earning potential, and to strike out on your own. It can be seen as the very underpinning of the American Dream; to rise from poverty to great wealth and power through one's own hard work and talent. It is why, so the theory goes, any American can be President - unlike the archaic Europeans with their monarchies and aristocracies. It is the inspiration for the fabled sense of pioneering spirit that led to the expansion of the country westwards and everything from the oil and gold rushes to the development of Silicon Valley.

Apart from a young, ambitious nation in search of a guiding philosophy and tapping into the zeitgeist, it is no surprise America adopted Smith's ideas. The nation was founded by brave, revolutionary (if somewhat misguided) people seeking liberty and a haven from authoritarian, elitist rule. In that respect the founders of the independent America were the natural heirs to the original Pilgrim Fathers, the first European settlers fleeing religious authority and discrimination.

Smith not only hailed the power and right to freedom of the individual, he also criticised the role of over-bearing rulers. America had undergone an anti-royalist revolution sparked by trade tariffs and taxes imposed by a distant, self-interested government. A bloody, complicated war had finally achieved a long-held desire for freedom. It was the freedom to create their own constitution and political system based on reason and equality. A freedom to trade with whomever, however, and in whatever they saw fit. That said, not every American held Smith's free-trade capitalist view to be the ideal. The first President, George Washington, was wary of international trade and signed an anti-free-trade bill. He felt trade's social effects were negative overall and doubted any real economic advantages of importing goods - although he was an army man, not an economist.

It is also worth noting that the apparently noble aims of the Thirteen Colonies seeking independence from Britain were only achieved with the sacrifice and decisive assistance of the French and Spanish. The Dutch also became involved and it is no coincidence that War of Independence or American Revolution involved the major European trading nations lining up against each other. Whilst undoubtedly some believed in the moral arguments for self-determination and egalitarianism, most were driven by more self-serving aims.

By the mid-19th century, European politics had to deal with the complexities of international trade combined with industrial and social intricacies and the immergence of stock markets. People no longer lived in small, self-contained villages, working the land according to a subsistence model. They dwelt in towns and cities with complicated interactions, interdependencies and structures. They were employed in factories and offices, in households and shops. Different levels of income and expenditure emerged to provide for a variety of needs and opportunities. Wealth, although still predominantly in the hands of a long-standing aristocratic elite, was increasingly controlled by a merchant class of factory owners, industrialists and investors. This new elite were also keen to wield political influence, and politicians increasingly came from a commercial background rather than the born-to-rule classes.

All of these elements, their planning, maximising and predicting their interplay had become vital to a nation's wealth and even its very existence. As such, understanding this interplay was also vital. Science had advanced so much that it made sense to make a rational analysis of economics. Mathematics, alongside philosophy, was applied to phenomena like supply and demand, production, and taxation within nations and markets. French academic Nicolas-François Canard was a pioneer in this field, not just examining the fairness and nature of taxation but also using complex algebra to explain his work.

Canard's work enabled his compatriot Antoine Augustin Cournot to go further in applying formulae and mathematics in his seminal economic works. Cournot also introduced ideas of probability; a measure of how likely an action was to lead to certain outcomes, thereby introducing risk analysis to economics. Another Frenchman, Jean Charles Léonard de Sismondi suggested the idea of an economic cycle of growth and decline over a period of years. This cycle could not be altered only prepared for; pre-empting what would happen across a range of economic concerns. British mathematician and philosopher Alfred Marshall adopted efforts to advance the maths used in economics whilst also bestowing a moral duty on the discipline to improve the lot of all in society. He also believed that it was vital that whilst increasing the level of maths in economics, it should not be done at the cost of making it incomprehensible to the non-economist.

Another English philosopher, Jeremy Bentham developed the concept of utilitarianism as a method of government to best deal with the complexities of social and economic interactions. Bentham proposed that law and government should have one primary concern above all others: that of 'maximising utility'. He defined utility as the sum total of the satisfaction derived from something minus the suffering caused in its realisation. This idea is often summed up as 'the most amount of good for the most amount of people' (a corruption of a quote from Bentham). This concept, along with Smith, heavily influenced economic and political thinking throughout the following century.

Utility attempted to value a product or service in a way beyond the fundamentals of what was spent on harvesting, shipping, processing and selling it. However, as satisfaction or personal benefit cannot be simply measured, and varies from consumer to consumer, economists looked to the rather blunt tool of the willingness to pay. Bentham also set out ideas around why and on what people will spend disposable income, and this came to be a measure of utility – how much is someone willing to pay for something compared to how much they own or earn.

Like Smith's economics, this measure again depends on the notion of the rational actor. Given a choice between purchases, the consumer will look at the utility, the benefit of both options and choose that most useful to them. It assumes that they will have all the information available and make the choice that is best for them in the long-term. That they will spend the appropriate amount according to their wealth for the best product available to them. That choice then has other economic implications related to the cost of manufacture, supply and demand.

Bentham felt it was the duty of rulers to ensure as many people as possible were as comfortable, as happy as possible. He held a disregard for religion for just this reason. He also believed in the primacy of the individual; that a person should be free and that they had inalienable rights and in this he opposed slavery, capital and corporal punishment. His progressive ideas were influential but the difficulty in measuring happiness meant that his concept of utility was reduced to base financial terms. Although ideas around measuring how happy a population was would return over 200 years later in the wake of the 2007-08 financial crisis.

An important influence on American constitutional and political thinking, Bentham's ideas also informed what some might consider contradictory national policies of welfare and of individual freedom and equal rights. He paved the way for John Stuart Mill whose *On Liberty* also heavily affected the young America. Mill insisted that any nation state was only as good as how it treated its people. A nation that puts the work of its people ahead of the intellectual and moral improvement of its people; a nation that tries to belittle or make its people docile and predictable, would accomplish little. The systems, structures, factories and offices created by belittling the workforce would eventually fall. Many thinkers and writers of the Enlightenment, and later the romantic era (a very conscious rejection of industrialisation) saw the importance of human worth, of ensuring people are valued more than their work or direct economic contribution.

In the real world, as opposed to the theoretical world of thinkers, government's principal role was by now one of maintaining and maximising economic prosperity. That meant commercial success. Ever more aspects of national life were seen through the lens of their economic impact. The effects of how wealth operated and could be managed by government meant political theory and economic theory remained intertwined. But economics was increasingly concerned with analysis, and on influencing and predicting growth. The philosophical aspects of economics, however, still counted, with moral arguments over the best, fairest way to distribute the wealth being foremost.

As well as inspiring the whole economic model of a new nation in America, Smith's work demonstrated the rules of what would become referred to as classical economics. It was ultimately theoretical though. It could not be proven experimentally, at least no one was willing to make the huge sacrifices required to do so. As such, compromised versions of Smith's ideas had to be adopted. The realities of government, people and the world at large created too many obstacles. The predictions he made turned out to be both true and false to differing degrees. His ideas around productivity and trade were, to a large extent, put into action across the world. It would not be until the Russian Revolution and the brutal implementation of Communist ideology that any serious attempts would be made to challenge them in a national, industrialised economy.

Smith provided a foundation for many economic, philosophical and political thinkers to come. Writers that expanded on his ideas, introduced nuance, tried to undermine them and developed mathematical models around them. The worldly realities of production, people, trade and government created new versions of Smith's work, some seeking to amend it in view of its practical shortcomings. Perhaps the most famous writer to follow Smith is one whose ideas would radically reshape the world in the 20th century, although not necessarily in ways he would have approved of.

The work of Karl Marx appears to be the opposite side of the economic coin to Smith. He saw a problem with the dominant notion of individual self-interest. He saw workers - the majority of the population - as being wholly exploited rather than being appreciated for their role in the economic system. They were not being fairly rewarded for their labour. The worker, on whose efforts the capitalist system rested, was paradoxically powerless to change this. Their work was what made the wealthy wealthier, despite Smith's assertion that wealth should be reinvested. The workers' situation, if it improved at all, did so very slowly, over decades. The elite's pursuit of wealth actively reduced the workers' influence. They were no longer rewarded for the work they did but were paid what the industrialist could get away with. A new economic model emerged. One in which the worker, the basis of a nation's wellbeing, owned an equal share of the capital and of the assets. Workers should own the means of production, as well has the goods produced.

Marx's work was a product of over half a century of industrialisation. Of urban society becoming the principal way of life. His chief collaborator, Friedrich Engles, inherited a factory in Manchester. Engles, however, was never really a company man and he gave a practical, first-hand insight into the nature and application of socialism and Communism.

Marx and Engel's *Communist Manifesto* and *Das Kapital* would see Marx elevated to the status of the most revolutionary philosopher-economist of the 19th century, and the most influential of the 20th. Unlike most others who had concerned themselves with the distribution of private wealth, Marx stated that those fundamentally responsible for the generation of wealth should reap the majority of its benefits. The owners of factories and banks, the ruling elites, should have no more than the worker, clerk or farmer, and they should all have plenty. Such radical notions, of course, found little favour with those with the most invested in the status quo. The work of Marx would, however, provide the moral, political, economic and intellectual artillery in battles between workers and ruling elites for decades to come. In this respect he would represent the most political form of a discipline – economics – that moved more and more towards attempts at dispassionate, rational analysis, models and prediction.

Despite the slow infiltration of socialism and concern for workers' rights into society and politics, Smith's concept of economic self-interest held almost exclusive sway over capitalist economies well into the 20th century. Political parties fought elections primarily on matters of personal economics. The electorate faced questions and arguments over what sort of government would mean lower taxes for you, more job opportunities for your family, a better future for your children. Economic competence and good stewardship became the key to power. Economic incompetence became the likeliest reason for being voted out, or in extreme cases, for popular revolt.

Societies and their rulers found ever more complex ways to balance religion and culture into a system that bound a population together. Concepts of freedom, fairness and equality were combined with attempts to keep order, reward power, subdue revolt, and generally navigate an impossible route to utopia. A utopia that somehow people have come to believe in. The result of this was that money, its accumulation and influence, became the one thing everyone could agree on.

Furthermore, they agreed that more money was almost always a good thing. However, no matter how much there was, there never seemed to be enough, so how it was distributed would become the new battleground.

The self-regulating, free-market economic system of profit and investment, on a local, national, or even an individual level, continued. Government could never truly let commerce run completely free, primarily because of the necessity of taxation, but it allowed it as much freedom as it could politically afford to. Self-interest at all levels seemed to be the best way to go.

Marx and those that agreed with him would essentially contradict Thomas Hobbes' and Adam Smith's assertion that people are inherently, inescapably self-interested. In fact they are all ultimately wrong. Humans are no more inherently self-interested capitalists than they are inherently Communists or socialists. Neither state is natural but are products of human thought and myriad influences that see us favour one state or the other. That both are possible, and indeed have been seen to be possible, means both are as natural and unnatural as each other.

Hobbes and Marx, Smith and Galiana were geniuses of their time, developing and formalising concepts in a way no one had done before. Their ideas inspired revolutions, constitutions, capitalism and nationhood, whether they meant to or not. Whilst Marx predicted an end to the unsustainable excesses of capitalism and Smith predicted the ultimate evolution of a free, unregulated market, both were undone by human nature. Workers did win more rights and a greater stake in the means of production, but they remained, for the most part, capitalists and self-interested. Or at least, the majority did as and when it suited them. Globalised markets became freer, but not completely unrestrained and the resulting social and employment problems led to more authoritarian governments. The people, ultimately, wanted things both ways – often demanding rules to be imposed on some, but not on them; expecting help, but not giving it.

The very religious Jean-Henri Dunant was the son of a Swiss businessman. He himself worked for a bank and, assigned to the French colonies, found himself appealing to the emperor Napoleon III for assistance in his business. In 1859 the Emperor was on the frontline of a war with Austria and Dunant decided to petition him in person. The journey took Dunant through a battlefield in what is now northern Italy. 23,000 men lay dead, dying or wounded on the fields of Solferino. Dunant pleaded with the local civilian population to help the soldiers of all sides and spent his own money on medical supplies. Within four years Dunant had overseen the first meeting of the International Committee of the Red Cross. Within five years he had played a leading role in establishing the First Geneva Convention. All of this to the detriment of his own business. He was declared bankrupt in 1868.

Whilst Dunant did not gain financially from his endeavors, it is, perhaps, too easy to be sentimental about him. His legacy lives on in the work the Red Cross and Red Crescent do and the thousands, perhaps millions who are alive today that would not otherwise have been. His name may be largely forgotten by most, but was his motivation entirely selfless? Certainly a religious conviction can be said to be never truly selfless as most religious beliefs, and Dunant was a committed Christian, reward the good not necessarily in earthly ways, but in the afterlife. No one can truly claim to know Dunant's motives, but the desire to build a legacy – something that remains long after the instigator is dead – is frequently thought as being akin to seeking immortality. So the insistence on some sort of self-interest may still hold true, but in ways that cannot be economically accounted for. As for economics, all the money Dunant raised and spent on medical supplies and staff and hospitals entered the economic system as well, just in a different way.

Where self-interest is concerned one question economists had to ask themselves was how charity could be explained. Why it was that not all workers were horribly exploited in a race to the bottom in pay and facilities. Many factory owners, often compelled by religious conviction as well as new perspectives on productivity, appeared to care much more for their employees. They provided housing, sanitation, education and healthcare as a means of motivating workers. A happy worker, one less exposed to illness or worry for their family, one housed nearby, would turn up on time, work hard, and be less likely to strike. When it came to skilled and semi-skilled workers and those higher up the scale of clerks and office workers, good, talented employees would also be loyal.

This move towards wealthy employers spending money on assets that did not directly produce a capital gain would come to highlight a flaw, or at least a further complexity, in the self-interest principle. Not every employer or politician agreed on the wisdom of such indulgences for workers, but even apparently altruistic actions have a selfish motivation. Building houses, infrastructure, and hospitals made sense to those seeking to keep workers healthy, loyal and close to the factory in order to get the best return on their wage bill. Building schools and training colleges meant a ready supply of capable workers.

Even the public image of an industrialist or their company can receive a boost from charitable work, making consumers more amenable to their product or service. But how to financially account for it and its influences on productivity and profit?

In Britain, the Cadbury family built an entire town to support their chocolate factory. Alcohol could not be sold in the area because of Cadbury's religious convictions. The fact that a sober workforce is probably a more reliable one is a happy coincidence. In America, Andrew Carnegie, steel magnate and one of the richest men that ever lived, built over 3,000 libraries across the US along with arts and educational establishments.

Carnegie belongs to a group of late 19th and early 20th century entrepreneurs known pejoratively as the robber barons. They made millions (billions in modern terms) in banking, natural resources and transportation and were known for their ruthless pursuit of personal wealth. Through his rise from poor immigrant to the richest man of his time Carnegie broke strikes and drove down wages, ironically keeping down those he would later claim to help. The philanthropic endeavours of the likes of Carnegie, banking magnate JP Morgan, oil tycoon John D Rockefeller and the other robber barons were dismissed by Theodore Roosevelt who said "no amount of charity in spending such fortunes can compensate in any way for the misconduct in acquiring of them."

It is widely believed that Carnegie, and the others of his era, were motivated by religious belief or particular notions of morality. He is quoted as saying that "the man who dies rich, dies disgraced." In seeking to redistribute or undo some of the ills done in the acquisition of their wealth the robber barons believed they would find some spiritual salvation. Whilst the word philanthropy is derived from a Greek word used to refer to a superior type of human, it came to mean a love of humanity usually exemplified by generosity. More recently it has been co-opted by the wealthy to refer to their own particular brand of largesse. Scientific research, libraries, campaigns for social justice, even infrastructure projects now rely on very wealthy people, and their sense of philanthropy, in order to be realised.

One of the major problems with philanthropy is that its greatest effects are often unintended and unpredictable ones. The philanthropist will invariably direct their efforts at something they personally feel is important, perhaps something that has directly affected them, or simply a misguided idea of what will improve the world. In isolation, suddenly improving, say, libraries, whilst the less attractive problem of sewage remains un-tackled does little but provide another building in which people can spread disease.

As well as undermining the rational self-interest tenet, one that, were it true, would have held back human social development by millennia, philanthropy also skewed the economics of a country. Government tax could be directed elsewhere, but with private philanthropic money taking up the shortfall, it becomes a lot harder to account for and to measure its effectiveness.

It would be government's role to take a dispassionate view of what was needed to support and create a healthy, productive, educated and happy society. It would have to do so by analysis of numbers, rather than the personal or emotional aspects more prevalent in charity. Today, charities aiming to increase donations will not use statistics about how many people are suffering, for how long, how many will die, the cost of vaccines and so on. They will show one emotive, heartfelt, specific example and then say many others suffer the same terrible fate.

The latter is the far more effective because it appeals to our irrational, impulsive nature. Rational thinking is more selfish; it prioritises self-interest, but if that is all humans considered charity would not exist.

Philanthropy reminds economics that morality is a key decision-making influence. If self-interest is rational, and therefore correct, there is no need for morality to enter into the argument. Science operated likewise; if something can be proved to be fact, morality had little impact. Religion had held ultimate sway over matters of society and morality as well as gifting power and insight to those in charge. But the rise of science and commerce in individual and collective wellbeing diminished the power of religion.

THE RATIONAL IRRATIONAL HUMAN

In the mid-to-late 1800s, Friedrich Nietzsche contributed to the ferment of revolutionary ideas that was ensuing across the Western world. He was another of the era's thinkers to consider the nature of humanity, morality and free will. He challenged the prevailing ideas around the equality of all people and what was and was not an admirable, desirable human virtue. He also questioned the dominance of religion, specifically the Judeo-Christian tradition, over morality. Whilst not necessarily attacking religion's achievements, he felt that religion had done its work in advancing humanity as far as it could and it was time for the next thing. Nietzsche's famous quote "Gott ist tot" (God is dead) was less an assertion of his own atheism and more a commentary on how the rapid developments in science in particular, but also in ideology, had 'killed off' the need for a God. He feared for what would take religion's place; what would fill the vacuum. It was a fear that would be justified as the rise of Nazism and Communism would both see spiritual religion effectively replaced by dangerous, modern nationalist-humanist quasi-religions.

Much of Nietzsche's work looked at the suffering humans were apparently destined to endure. Much of that suffering would be born of the desire for things – in particular status and wealth. His notion of the 'will to power' was an early take on what drove the human spirit and mind. Humanity's drive towards self-conservation led it to impose itself on the natural world. Counter to Utilitarianism, Nietzsche claimed humans were not striving to find the most amount of happiness, rather that they simply sought to manage as effectively as possible the suffering they inevitably has to endure. It was a subtle, yet important distinction.

In order to achieve this tolerance of human suffering Nietzsche proposed the idea of the 'ubermensche'; the ideal human to which we could aspire. It was an idea that would become much corrupted, largely due to his sister's support for Nazi ideology and her role in managing Nietzsche's estate after his death. The ubermensche or superman was a thought experiment. The hypothesis was that a 'perfect' person, one free of suffering, could only exist in isolation from the peer pressure, social mores, imposed morality, and constrictive thoughts of others. Free from these constrictions, humans could be free to achieve their true potential.

Peer pressure, morality, suffering, and free will all accounted for human behaviour in new but hard to statistically analyse ways. By the close of the 19th century, rule by faith and wisdom was increasingly being overtaken by rule by numbers and the rational. Those numbers could be voter numbers, campaigning funds, tax income, life expectancy, or the number of people in education. Many believed that the person or nation with the best numbers was winning the race. As people became better educated and informed, it also became necessary for rulers to demonstrate in simple, irrefutable ways that a decision or course of action was the right one.

It was clear that many aspects of human behaviour were not simply governed by the accumulation of personal wealth in order to improve their material circumstances. Other elements came into play when attempting to understand or predict their economic activity. Empathy, morality, trust, and an assortment of personality traits - these were aspects of every human, rich or poor, powerful or lowly, that meant they could not be relied upon to coldly examine every exchange for individual economic advantage. They would not always be rational and predictable. Worse than not knowing whether every person would think in this way was the fact that sometimes they might, and sometimes they might not.

By the dawn of the 20th century economists had largely written this behavioural anomaly off as freakish; small enough as to be dismissed without dramatically affecting their central assertions. They maintained the self-interest concept at the heart of their work. Economic analysis and prediction was complex enough, without trying to incorporate morality and peer-pressure. The claim was that people still acted out of rational self-interest and if something in the economic process could not be represented numerically, it did not have a significant impact.

As the complexity of what economics attempted to measure, explain and predict increased, so did the language economists used. The study of economics, the implementation of that learning, and attempts to test new economic theories occupied great minds. The influence of economics as a discipline was felt throughout society.

As economic models evolved, and the store of historical data built up, so more and more patterns were found in more specific areas. Where obvious patterns could not be found, ever more complicated mathematical models incorporating ever more influences, survey results and market data were developed so that patterns would, eventually, be found. Patterns that would predict future economic events.

The rise of science as the saviour of lives and economies saw its methods applied, often incorrectly, to more aspects of the world. Its success at explaining and analysing meant it found a welcome home in economics. A subject that had its roots in philosophy and politics would become ruled by the movement of numbers, prices, quantities and volumes. So logically it could be explained by mathematics, proofs and formulae.

Statistical analysis and probability became foremost aspects of the economist's work. It enabled models to be created; something that produced results, information that could be used for decision-making, rather than more abstract, philosophical dialogues that were equivocal and inconclusive.

Everything acquired a numerical value; a measure. Everything could and should be reduced to data so that it can be processed with formulae. Definitive results could be derived to predict the effects of policy change, population shifts or increases in supply and demand. People, labour, productivity, investment, the value of a day's work in a coal mine compared to a day's work in an insurance office - it all had a number. Numbers were clean, unarguable; they provided a clear indication of success or failure and what trajectory a nation was on. Nowhere was the cutting-edge methodology of numerical analysis more welcomed that in the new, innovative, capitalist nation of America.

10

ECONOMICS, POLITICS AND AMERICA

> *"I believe in America. America made my fortune."* - **'Amerigo Bonasera'**, *The Godfather*

America was a country 'discovered' by Europeans seeking new trade routes and new goods with which to trade. It was first ruled by Europeans who sought only to extract its wealth. It was then colonised by those escaping marginalisation and oppression in Europe. America's population grew as thousands, then millions sought new lives and new opportunities. As an independent nation it was founded, in part at least, as a reaction to self-serving, overbearing British rules on trade. A vast continent, it was slowly opened up by different groups of settlers, all seeking their own fortune. Huge reserves of natural resources combined with a pioneering, entrepreneurial spirit to create a superpower and a vision of human achievement.

Since its European discovery in the late 1400s and through to the 20th century, America had abundant resources to enable it to both trade internationally and to be self-sufficient. It may have needed industrialisation and a reliance on slavery to realise it, but it had the power to export without necessarily having to import.

Britain, on the other hand, had taken the route to becoming a trading nation. By the 1800s it was heavily reliant on the importing of food, oil and other essentials. In order to make the most of international trade (as Adam Smith had suggested) it had chosen (by accident and by the design of its political rulers) to specialise in manufactured goods. Deriving a high profit by buying cheap commodities and selling high quality, high cost manufactured goods to customers all over the world. Britain was not alone; many European nations had done the same, including Germany.

Political, economic and social pressures in central and Western Europe saw the Austro-Hungarian Empire drag the continent towards war. Powerful trading blocs and monarchical families across Europe took sides. Britain, France and Russia united against the Germans and Prussians in a desire to protect vital continental and Mediterranean trade routes. That in turn inflated local political pressures into the global conflict of the First World War.

Trade and industrialisation had disrupted the long-settled world order and created international tensions and insecurities. Rather than address them, those tensions had been allowed to foment to uncontrollable levels. Politicians took a risk in allowing war to become all but inevitable; some privately believing that a war would resolve many long-standing issues of political and commercial influence once and for all. Even the most cynical politician, however, would almost certainly have thought twice had they realised the scale of the slaughter that the so-called Great War would inflict on ordinary people from Australia to Canada, India to Ireland. Not only was the War fought because of industrialisation, it was fought with industrial methods and machines.

Politically, America attempted to remain independent in the First World War, even after they had aligned themselves with the allied forces of Britain, France et al. It would not be the only time that the fiercely independent, self-reliant US nation would prefer insularity to interfering on the world stage.

At the end of the War, many felt measures needed to be taken to prevent such an atrocity happening again. They still believed that war in 1914 had been inevitable in the circumstances, but a repeat of those circumstances needed to be foreseen and avoided. Some form of international agreement and regulation was needed. A body that would prevent such wars and diplomatically find common ground and passive resolution to disputes. Not every nation was in a position to establish a new global order, and in some cases they did not really want to. A jockeying for position on the global stage started to emerge.

The end of the War in 1918 left a fragmented Europe, a post-revolutionary Russia, and an increasingly isolationist America to reshape the world. Only the US economy, however, was not significantly damaged in the years after the war, but its refusal to play a leading role in rebuilding global trade brought about a slump. Within a decade, the Wall Street Crash had led to the Great Depression which echoed across the political west. Over the course of the 1930s things deteriorated.

Although the US started to reverse its isolationist,

protectionist trade policies under President Roosevelt, tensions, particularly in Europe, only increased. The dire economic problems of Germany combined with a sense of popular post-war injustice. The wartime allies may have disagreed, but Germany was not officially defeated- she had agreed to stop the fighting only to be treated politically as a defeated nation. The resultant social division and political strife facilitated the rise of Nazism and ultimately the Second World War.

By 1945 the world had borne witness to a turbulent, catastrophic 30-year period, caused ultimately by commerce, trade and its newly dominant role in society and government. Humans, with our short lifespans, struggle to see history in the context of centuries and millennia. The connecting together of the world, after having spent millennia being isolated from one another by geography, happened so quickly that it should not be surprising that it took decades to try to sort out the result. This chaotic, painful start to the 20th century, littered with human errors of judgment and selfishness, is now a lesson in what can result from the rapid, uncontrolled expansion of international trade and personal enrichment. Whatever conclusions can be drawn about what came after, the period of 1900-1950 in historical terms was the growing pains of a new world order. The inevitable bursting of a dam which had held back the economic and political waters built up by creeping globalisation over the centuries.

Once the dam had burst and the waters drained out, a strange sort of peace befell the world, ushering in a new era for global society. With Russia, now the imperial core of the USSR, concerned with its own post-war political expansion, and Europe shattered physically and psychically, America stepped in as chief lawmaker and enforcer of global trade. It saw both an opportunity and an obligation to fill the power vacuum before someone else did. Having already laid the groundwork before the Second World War, the US led an international coalition to form some sort of broad, mutually agreed system of global free trade. This coalition would see some nations join enthusiastically, some feel pressured into joining, and some vociferously resisting.

By the early 1950s it was evident that the US and its allies would not be doing business with the new USSR and Communist bloc any time soon. Free trade became the basis for an alliance of European and Asian states lined up against Soviet influence in the world. Essentially two separate blocs, both doing economics and politics very differently, formed. With little appetite for military conflict the prevailing mood was that, as long as one did not interfere too much with the other, there would be no major problems.

It seems hard to imagine now that the US and USSR were not from the outset bitter, implacable enemies throughout almost all of the 20th century. The union of the US, Britain and Russia during the Second World War is thought of as a coalition of convenience; a necessary evil. Stalin could not be trusted in peacetime; after all, he had thrown his lot in with Hitler early on in the war, even if it was somewhat equivocal. Yet the two sides of what would become the Cold War were not always so diametrically opposed and openly suspicious of each other.

The Truman Doctrine, President Harry S Truman's 1947 foreign policy declaration, stated that the US would oppose Soviet expansion in all circumstances. It stopped short of military support for those nations threated by Stalin's growing desire for dominance of the world. However within two years of the statement NATO, the North Atlantic Treat Organisation, established a military alliance of Western Europe, the US and Canada, and other friendly states, which would impose both a military and economic opposition to the USSR and its allies. The US became the chief anti-Communist police officer in former European colonies from Chile to Angola to Vietnam. Under the influence of the US these countries had, for the most part, been vacated by their former imperial overseers in the post-war shake up of the world. Those former empire powers had deliberately left behind an assortment of corrupt, autocratic, puppet rulers. Rulers that were by their nature weak, given to corrupt dictatorship, and therefore liable to popular revolution. Those revolutions provided an opportunity for Communist expansion.

The history of the Cold War is long and complex, taking in the military, economic, political, philosophical and even technological. It is seen in many ways as a dark period for the world, one where the very real threat of nuclear Armageddon loomed almost daily for nigh on 40 years. Yet like so many of the revolutions that have changed the world, it had unintended, unforeseen consequences the effects of which lasted even longer.

The Space Race, the competition between the USSR and US to control space, saw incredible human achievements and advances in technology unimaginable in conventional political-economic conditions. Whilst nuclear weapons threatened the world, nuclear power changed innumerable lives and the physics behind it radically advanced understanding of the universe. These two connected fields would directly feed the digital technology of today that would revolutionise the global economy once again.

In those early, frosty days of the Cold War, US political

leaders posed a question to leading academics, including those of the economics department at the University of Chicago. What does the political West believe?

In the USSR, that inscrutable accumulation of Communist nations now dominating Europe, they had a guiding philosophy. Marxism, a tough, modern economic and political theory, was being put into practice. The corrupted nature of its implementation was open to very serious question, but nonetheless, there was a core idea that people could easily grasp. The problem was that the USSR was now the enemy. Not an enemy, as had been encountered in the past, that was seeking commercial and imperial advantage, but an enemy of ideology. It was an ideology that many people felt was not without its merits.

Marxism stated that the worker – and as such the vast majority of the population - should share equally in the product of their toils. It made clear that the employer, the boss, was the exploiter-in-chief, growing rich and fat and lazy off the back of hard, honest work. Worse than that, some became wealthy without any work being done at all. The ordinary person should not be governed by some self-serving elite that controls the money, the levers of power, the laws and the property. Marxism sought an end to this Industrial Revolution injustice. The worker should be rewarded fairly. The boss should be brought to book. Work should be for the benefit of all. There should be equality.

What made this situation worse was the success of Russia in the Second World War. The Communist nation were the first into Berlin. They held back the German onslaught and crushed them utterly. They suffered greater losses at the front than the Western Allies, yet bore that burden and emerged victorious.

Although facts would later come to light that would cast Russia's military and broader domestic strategies as dubious to say the least, their centrally planned, bureaucratic, workers' nation had done at least as much, perhaps more, than America had to win the savage war in Europe. The Soviets also engaged in a propaganda war throughout the 1940s and 50s, meaning it was hard for outsiders to see anything other than a broadly positive, or at least, successful view of the USSR project.

Many would see University of Chicago economist Milton Friedman as the modern-day heir to Adam Smith. A man who would revolutionise post-war economics with new ideas and analyses. For better or worse, he would also become a talisman for the political right with his vehement support for free markets and for limiting a government's economic role to controlling money supplies and little else. That aside, Friedman's belief in the power of markets and the necessity of small government was profound. As such, Communism was anathema to him and many of his peers.

The US had a problem. Whilst they led opposition to Soviet expansion politically and economically, their grasp on popular, social opposition was more tenuous. That was true to a degree at home, but even more so in the former imperial colonies that America hoped to influence. What the US and their wider allies needed was a philosophy to oppose the apparent success of Marxist-Communism. Something they could re-build a nation on; something that could offer hope, direction, and goals.

Until this point capitalism was little more than an economic system that had evolved over the last two centuries. It had been largely left to its own devises, with new ways of doing business appearing and disappearing every so often.

There was no central guiding text or philosophy unpinning it; just various economic analyses and theories that had developed and subsequently been adopted by political groups of assorted, slightly differing ideas and constituencies. There was nothing the ordinary working person could identify with; no badge they could wear or flag to pledge allegiance to.

With little but the benefits of trade and economics to unite the non-Communist countries of the world, economics was looked to for assistance in making capitalism an ideology rather than just a vast, complex process. Friedman tied together concepts of political, economic and personal freedom to produce a libertarian ideal that was uniquely American.

Drawing on the founding of the American nation, in parallel with Smith's *The Wealth of Nations*, Friedman aimed to counter the growing narrative of strength through social unity. America had come together to win the war, but now that was over it must return to a focus on individual achievement. Only by advancing personal achievements can a nation of individuals succeed.

The 1950s is now seen as the golden age for America; a period where a young nation, having dealt with war at home and abroad, now looked to defining an era of comfort, authority and confidence. Aspects of morality and social cohesion aside, economically, overall increases in disposable income drove increases in the consumption of domestic products, which in turn fed consumer wealth and so on.

For Friedman this was also a time to further debunk notions of why people spent money. Again, Smith inspired his harder, more modern thesis. As Friedman's economic forebears has also asserted, consumers, the free citizens of the modern world were rational actors, not emotional. The modern consumer spent when they could afford it on thing they needed to improve their lives, and not due to illogical reactions to outside events.

Friedman and his peers at the University of Chicago developed the Rational Choice Theory. Building on Smith's ideas, it set in stone the idea of 'homo economicus' – a stable actor of sound mind and with access to all the relevant information in order to make a rational decision that maximises their own personal economic gain regardless of any other effects it might have. This idea would sweep through economics, and by extension politics, with speed and certainty. Economics being so closely related to politics it is no surprise that this principle influenced not just economic policy, but military, diplomatic and social policy as well. Almost all government action was based on the sound, rational self-interest of the subjects involved, whether they were a Soviet politician, a developing nation state, a real estate company or a car buyer.

Whilst it might seem flawed and one-dimensional to modern eyes, Friedman's reduction of economic decision-making into Rational Choice Theory did at least demonstrate one sound concept: that economic theories should be as simple as possible. Although in that regard he was not as influential in his field as he might have hoped.

Friedman's faith in the individual and his determination to reduce government's role in economic affairs would have a much greater long-term effect. Whilst his voice outside economics would fade somewhat in the 1960s and 70s, it would return even stronger in the 1980s and the Presidency of Ronald Reagan. Regulation on commerce was stripped back, as was welfare support and the taxation that funded it. Friedman's ideas also gained traction with the World Bank and International

Monetary Fund, no more obviously that in those intuitions' roles in designing the economies of the former Soviet states at the end of the Cold War. The consequences of imposing a free market, small government, light regulation model, which had been allowed to develop gradually, if not always trouble-free, in the West over the decades, are increasingly seen as seriously ill considered.

When the Cold War still raged, however, the University of Chicago gave birth to another idea fundamental to American anti-Marxist philosophy. Friedman's colleague Gary Becker, along with Columbia University's Jacob Mincer (a labour economist), distilled and updated Adam Smith's notion of Human Capital. Whilst Marx famously advocated the workers owning the means of production, Becker and Mincer viewed the individual themselves as the ultimate means of production. Marx had viewed the factory or the farm as the means of production to be collectively owned. The Americans thought beyond that. Factories crumble, farms turn to dust, especially during an economic depression, but the worker remains.

Smith and subsequent thinkers identified factories and facilitates as assets in the economy. With capital invested correctly, these assets could become more efficient and more profitable, providing more money to invest further in more jobs and so on. Becker and Mincer viewed the human as the principle asset to be invested in. That investment, that improvement to their wealth generating potential, was their own responsibility, just as it was a factory owner's responsibility to invest in their machines. In this case, however, rather than spending on research into better equipment, the individual human investment accounts for all of the education, skill, knowledge, talents, health, personality and experience they acquire over time. This individual investment in their own Human Capital is ultimately what creates wealth.

In many respects this idea goes back as far as Hesiod's *Works and Days* in which the poet reflected on how one could be a better worker, and as such more productive and successful based on individual experience and learning. In *The Wealth of Nations*, Smith had stated that the true potential of one's wealth could only be realised through trade with as wide a market as possible. This encourages specialisation, the skills in which will be enhanced the more one earns from trading with more and more markets. He defines human capital as the skill and work inherent in a worker. However, where Smith largely concentrated on the notion of labour, production and trade on a national level, Becker and Mincer's Human Capital reflected what this meant if you considered the individual as a sole, sovereign entity.

Communism had at its centre, regardless of how closely the Soviet experiment managed to actually stick to it, Marx's idea of the workers owning the means of production and sharing in its output. The principle of Human Capital countered this by suggesting that all free people - that is, those living under advanced, capitalist democracies - already owned the means of production. It was they themselves. Whilst under Communism it was the worker's duty to toil in their role for the benefit of everyone directly, under capitalism it is their duty to invest in themselves; to work for themselves. Flexible and resilient, the individual was an asset to the economy as much as any factory or farm. It was for each individual to invest time, effort and money in themselves to maximise their earning potential.

Friedman took on Human Capital as the basis of a contract between the individual and the nation state or their employer. This would economically account for the money spent training or educating them, and the return those in charge would potentially earn from that investment. A person is now an economic asset to the country. An asset that could either create money or cost money. It was now the patriotic duty of the individual to create money, to invest in themselves, to work for themselves, and to achieve, and in doing so their nation will thrive. Turning people into assets also had the by-product of creating a market in people, with companies and nations bidding for the best (most skilled, hardest working, cheapest) available.

The individual versus the collective. The free versus the beholden. It came to define the Cold War, and perhaps more than any other nation it came to define the US – the land of the free. It still does. The American Dream; that anyone, no matter how humble their origins, can rise to whatever they want – glory, wealth, power, happiness – if they just work at it.

The obvious downside – a pernicious, damaging downside – of this philosophy is that should you fail in your goals of glory, wealth or power, or even just a comfortable standard of living, then the blame lies squarely with you. There is no allowance for luck, circumstances, environment or upbringing. It is based on the notion that any challenge can be overcome, no matter how stubborn or severe. It is based on the notion that every achievement is open to anyone. The practical reality is that it is simply not true. If it were, there would be many more hard-working people from poor backgrounds in positions of political, cultural and commercial influence in the West. There would be more lazy, rich people living in social housing.

Economics would be proven to be both right and wrong.

Appealing to a self-centred view of humanity would frequently work for many in the population, although that might differ depending on the population being analysed. Equally, many would reject this view for a myriad of different, often irrational reasons. More often, people would act in both ways and governments and economists would no more know why and when than rulers and their priests of centuries past would know why God did what He did. They would attempt to go with the majority, but that would not always be clear, and would leave a difficult, disenfranchised group out in the cold.

As the world became more complex, and the humans that populated it open to ever more influences, it became necessary for governments to plan ever more carefully. With the economy the dominant factor in national, and during the Cold War, supranational success, the economic forecast became vital. But it was not alone. Other aspects of the world needed predicting and the rational power of economics spread throughout government and commerce.

PART 3

BEYOND ECONOMICS

Pure economics did not account for everything. It was complex, and would become ever more so, but it relied on something even more complex – humans. In order to plan, in order to see what the future might hold in a globally inter-dependent world, it would be necessarily to predict human decision-making, both collectively and as individuals.

This need would see two key areas grow in importance: psychology and data. The former would explain why, despite much examination, humans still appeared to act unpredictably, especially when money was involved. The latter would, it was assumed, hold the key to human behaviour – gather enough data on a person, and unlock the pattern to who they are and what they'll do.

Human data would be simple enough, gathered via surveys and records, but it would take a dramatic, rapid turn as technology moved to the fore. It was not enough for technology to process data gathered from human sources, it had to disintermediate. If human interactions and activities were increasingly open to measurement, predictions might finally be both accurate and universal.

11

PREDICTING BEYOND ECONOMICS

"Foreknowledge of the future makes it possible to manipulate both enemies and supporters."
   - Raymond Aron, *The Opium of the Intellectuals*

From the ultimate nature of the economy and society like Smith and Marx, to a detailed analysis of inflation rates, trade surpluses and currency values, economists moved from the very big to the very detailed. The complexities of 20th century economies and the increasingly huge, slow-reacting businesses that underpinned them, demanded planning and preparation. Understanding what will happen could be the difference between prosperity and poverty, or for a democratic government, being in power or in opposition.

Having looked to physics and science for inspiration to create models of economic behaviour, the latter increasingly involved itself with ever more complex predictions of ever-finer details. Individual companies as well as governments looked to economists to inform their strategy, globally, domestically, even within local regions. Economists started to specialise in industry sectors, in specifics like wages, property or retail. Economics as a discipline took on a life of its own, such was the demand for its knowledge and foresight.

Whilst governments still obsessed with economics as the foundation of their national wellbeing, they wanted other perspectives on the future too. They wanted to know what their strategies should be beyond the purely economic. How should they prepare militarily or socially, in areas of infrastructure or education, healthcare or technology? How could they be sure, or as sure as possible, their policies were solid, reliable, and the best solution for the years ahead?

In the aftermath of the Second World War the world faced what seemed to be a new set of rules. People and government struggled with the horrors of a war that affected so many civilians as well as soldiers and destroyed so many towns and cities. Leaders revised their long-held assumptions in light of the carving up of what had been the world's most politically, ideologically and physically powerful continent. The holocaust, nuclear weapons, American supremacy, new technologies, women in the workplace, welfare, national debt – a terrifying to-do list faced modern politicians and their advisors. Many felt the need to try to prepare for a future which had never seemed more uncertain. Others felt the need to imagine, and to build, a better future on the rubble of the old certainties.

If market economics has taught the world anything, it is that where there is a need, someone will try to fulfill it. The need to understand this new political and social dynamic inspired the study of the future. Applying multidisciplinary academic methodology akin to that of areas of history, universities started producing students and experts in what the future will look like; what patterns in the past can be applied and extrapolated for today.

By the 1950s, the serious, academic study of the future, previously a high-minded but indulgent preoccupation, had become important with the rise of the USSR as a nuclear power. Strategies were formed. Scenarios were built. What would a government do in the event of half the population dying in a nuclear attack? Who would most likely die and who would survive? What would that mean? How could a country be rebuilt? Naturally the US, the prime target of such any such attack, took a lead in this area of study.

During the Second World War, and for the first time in conflict, numbers played a vital role. The period 1939-45 saw many incredible and terrible developments, but one of the less famous was the rise of war by numbers. There was a new role for mathematics: as a weapon of war and as a means of analysing the success or failure of a strategy.

In Europe, the Germans and British fought a now famous battle of code-breaking and encryption. The Allied efforts were centered on Bletchley Park, the mathematician Alan Turing and his colleagues, and the nascent computing array known as Colossus. The Bletchley team created formulae of such complexity, and that required so much data, that whole new systems - of mathematics, of machinery - had to be invented.

For the US Army, in order to minimise risk and the potential for human error, field combat was increasingly a scientific matter. With ever more powerful and potentially accurate weapons, it was no longer a case of pointing at the target, firing and hoping for the best. Science took over some of what was once the sole concern of human instinct or skill. US artillery required extensive calibration charts. These charts indicated where to aim for given the variables of ammunition weight, weather and climate, geography and so on. Those in the field needed the best possible chance of getting a result from firing a weapon. In order to do so, numbers were calculated thousands of miles from the arena of combat, in offices in America, by hundreds of women. They became known as 'computers' and they worked on calculations for field artillery, aircraft and other areas of military research. Their groundbreaking work, done manually and with mechanical calculators, led to electronic computing and a drive towards statistical analysis of real world variables and outcomes.

After the War the US tended towards a technical, analytic approach to future planning. In Europe, including the USSR, it was more focused on the long-term outlook for societies, the environment and abstract ideas of language and thought. Over time scientists and inventors started to collaborate with political thinkers, economists and other social scientists such as psychologists, historians and sociologists.

Government and business started to put ever more emphasis on what the future would look like and what their role in it would or could be. They would, inevitably, seek to put themselves at the centre of any portrait of the future. As such those groups of researchers and thinkers that could put those organisations that were commissioning the forecast at the centre of a predicted future, reassuring them of their relevance, were the ones that were in demand. Consequently, those groups became part of an established ruling elite that would influence the very future they were employed to objectively predict.

In Europe, one such predictive organisation was The Club of Rome. The Club drew together former politicians and public servants, businesspeople, scientists and economists "who share a common concern for the future of humanity and strive to make a difference".

In 1972, The Club of Rome published its first report. It was a prediction, based on computer simulations, which stated the current rate of natural resource usage would bring about an end to global economic growth. That alarming warning was followed two years later by the publication of The Club's Mankind at the Turning Point. This time the report incorporated the regional and the social as well as technical data into a model that used 200,000 equations to conclude that, whilst catastrophe was on its way, it was not inevitable, and that human intervention could avert disaster. Fortunately, that was the limit of the Club's remit; designing and implementing that intervention was the responsibility of others.

Another, more influential, organisation was founded in America 25 years earlier. The Douglas Aircraft Company was founded in 1921 and by the time America entered the Second World War it was already a key supplier to the US military. Its business flourished in the early days of commercial flight with its DC series of planes, and between 1942 and 1945 the company produced 30,000 aircraft for the armed forces. The fifth largest recipient of government contracts during the War, the end of conflict saw the company facing serious financial challenges. The military decided to charge Douglas with a new role; one that had never really been considered before. Douglas would take many of their thinkers and engineers; those whose understanding of physics and complex mathematical analysis had helped produce not just brilliant aircraft and weapons, but had solved big strategic problems, and assign them to Project RAND.

RAND, a name derived from Research And Development, was charged with preparing the US for an uncertain future. A future where global nuclear conflict was possible, perhaps even likely, and in which science and technology would have an important role in the outcome. RAND would be a bridge between military planning and government policy making. It would look at the economic, social, political and military factors and develop robust, comprehensive plans and programmes.

Eventually, due to conflicts of interest between what Douglas manufactured and the issues it was advising on, RAND was separated from its parent company. It became independent and non-profit, predominantly funded by the foundation run by Henry Ford's family. It provided in-depth analysis for the US military. Using the expertise of scientists, economists, engineers, psychologists and military figures it built scenarios, calculated probabilities, assigned numerical assessments of risk and considered the options for responding and their consequences.

In the Cold War years of Soviet threat, both real and perceived, much of RAND's resources were focused on the likelihood of nuclear war. It famously developed the MAD (Mutually Assured Destruction) doctrine which asserted that neither actor in a potential nuclear conflict would strike first as to do so would ensure the ultimate obliteration of both attacker and target.

Amongst many other theories RAND developed to inform policy and strategy decisions was work on Game Theory. It was Game Theory that influenced the MAD doctrine as well as policy on the nuclear arms race and the Space Race.

Game Theory is a multifaceted set of ideas that, broadly speaking, propose that humans will frequently act in ways that are not to their advantage but do so for reasons we can all understand - that is, assuming we are not psychopaths. Although applied in many situations, financial transactions are the most readily used illustrations.

One common example uses the notion of fairness. If person A is given £10, but only on the condition they share it with person B and that person B accepts the deal, both get to keep the money. If A gives B £1, B will reject the offer, despite it being free money, and neither will get anything. B sees the deal as unfair and is willing to turn down free money to teach A a lesson. Experiments suggest that B will continue to reject any offer, more often than not, until the money is shared equally. B is willing to harm their own interests for the sake of a principle with no logical basis. Game Theory expanded various ideas around opposing groups acting against their own long-term interests if one side was seen to take an advantage in some field or another. What was true of research subjects in university would be true in boardrooms and government offices.

RAND's work applying ideas like Game Theory stretched to take in areas of American life including welfare, health and the use of digital and computer technologies. RAND sought to reduce everything to cold logic, even the illogical reactions set out in Game Theory. A utilitarian form of thinking designed to take emotion out of key decision-making. Like the modern economist, RAND would predict enemy actions, and therefore inform potential American reactions. They assumed that any opponent would always act in a manner that could be rationally assessed and predicted.

RAND became something of a target for satire, its

shadowy, unaccountable influence, particularly during the Cold War a source of conspiracy theories. It would also become well known because of one of its strategists in particular, Herman Kahn. Kahn's ideas around how the US could win a nuclear war were controversial at best - as one might expect from any policy that allowed for an 'acceptable' number of causalities - but he found wider fame as chief inspiration for Dr Strangelove, the eponymous character of the Stanley Kubrick film, played by Peter Sellers.

Having built a reputation for thinking the unthinkable, RAND's work moved into education, justice and welfare. During the Vietnam War RAND had been involved in the controversial 'body count' measure of the War's success. Without a conventional frontline that would advance into enemy territory, it was decided that the only logical way to track US and allied forces' progress was how many enemies were being killed. With little or no appreciation for the realities of that particular conflict, the results were horrific. Soldiers were tacitly pressured into shooting first and asking no questions later. It was a policy which only reinforced North Vietnamese resistance and made the War arguably America's worst foreign policy action in its history.

RAND's scientific and engineering background perhaps explained their fondness for blunt, statistical analysis. It started to reduce more and more aspects of the world to data-driven, analytical solutions. This would see RAND develop and increase in influence as computing science advanced. Ever greater processing power was required in order to build scenarios, estimate probabilities and create policy options of unimpeachable mathematical logic.

RAND also played a key role in the fundamentals of

satellite communications and the internet. Originally conceived as a way of maintaining communications in the event of a nuclear strike, telephone company AT&T refused to be a part of the internet, leaving the Pentagon to take it up. As a result, government, not a commercial company, laid the foundations for the world wide web and, arguably, ensured its open access to all. RAND would also employ an array of lauded economists, mathematicians and political figures including numerous future Secretaries of State and Defence, and also many other powerfully influential figures. It would become an established part of US policy making and advice.

A problem remained for government and business. Prediction remained notoriously hard to get right, especially when it really mattered. Even with all of this data and clever people and expensive computers that RAND and their ilk gathered together. What made it tougher was the timeframe. Planning 10, 20 or 50 years ahead had serious limitations. The Communist bloc had their own long-term plans based around production and social order. A regime which saw itself as immutable for decades into the future had no problem announcing what was required for the future success of the state. The problem was what they were willing to do to achieve those aims (or for the appearance of achieving them).

In the democratic West, however, governments rarely lasted beyond five years, and certainly not more than ten. Forecasts looking at a world in fifteen or fifty years held little policy sway beyond rhetoric, usually around election time. Plans for schools or energy or healthcare or even the military often needed generational timeframes. But elections demanded popular, short-term, quickly implemented ideas. Voters could not be trusted to think long-term; they seemed to assume that everything would just work out.

In a world of ever-advancing science, technological progress, growing populations, international trade, and new products, business also had to find a way to deal with this complexity. The pressure to make ever more profit saw them take ever-greater risks. But no sensible business wants to take a leap into the unknown. Business was also coming to terms with the risk of not getting the future right. Business history is littered with examples of poor judgements that could have finished the company off. The Western Union memo that said telephones were of no value to the company. The Warner brothers predicting little public appetite for talking pictures. Yet getting these predictions right was increasing not just a matter for the future survival of the business, but also of its current success.

WHEN ECONOMICS FAILS TO HELP

Less than 100 years after the Founding Fathers created the modern American nation, the robber barons built it. Even in the age of science and reason, however, power did not always stick with the rational. Those gifted power, ultimately, by science did not necessarily see logic as the solution to everything. Certainly not in their desire for prediction.

The term robber baron was originally coined in Germany in 1810, but it referred to a breed of corrupt feudal lords that reigned many centuries earlier. Lords that stole from others whilst exploiting the protection their legal and political status afforded them. It is a term now more readily associated with the likes of Andrew Carnegie, Jay Gould, Cornelius Vanderbilt,

John D Rockefeller, and John Pierpont (JP) Morgan.

These entrepreneurs, whose lives spanned the late 1800s to the early 1900s, substantially defined America, the American Dream, and the collective view of modern American commerce. Their lives were by parts inspirational and cautionary. They were often religious men, indulging in philanthropy for influence and power, but also for spiritual reasons. They were amongst the richest men in the world, and invariably the most ruthless. Pioneers of their time, they exploited opportunities, established companies, built buildings and created legacies that exist a century later. There was, however, one thing that their vast wealth and power could not buy – clairvoyance.

Morgan is quoted as saying "millionaires don't need astrologers; billionaires do." Amongst many other interests, he had a financial stake in the White Star Line, the operators of the Titanic. He was rumoured to have cancelled a trip on the ill-fated ship's maiden journey at short-notice and on the advice of his astrologer. Although his astrologer did not tell him to divest himself of his stake in the company that was ruined by the disaster.

Charles Schwab trained as an engineer and joined Carnegie's steel company. He rose through the ranks to become president by the age of 35. He and Morgan arranged the secret sale of the company to create the behemoth US Steel, which Schwab also headed. The pair also shared a key advisor.

Evangeline Adams was a contemporary of both Morgan and Schwab, but she did not work in commerce. She was an astrologer. In 1914 she was tried in New York for fortune telling, akin to fraud at the time. She was acquitted, with the judge having been convinced that hers was a study of science because of the maths involved in her predictions. Maths has been used to track and accurately predict the movement of the stars for centuries, but it had never managed to predict the actions of humans.

Adams was said to have predicted both the Wall Street Crash and the Second World War, which strengthened her reputation as a predictor of uncanny skill. Charles Schwab died heavily indebted, having lost huge sums in the Crash.

Psychics have been consulted by US Presidents from Washington to Reagan via Kennedy. One of Reagan's Chiefs of Staff even went on to say that virtually every major decision during his time in the White House was cleared by the Californian astrologer Joan Quigley, who would draw up horoscopes to assess the chances of an enterprise.

Each of these powerful men was desperate for reassurance beyond the wisdom of humans that their decision-making was correct. Each sought reassurance from something 'other' when logic and data failed to make the grade. They had access to reason, and all three were religious men, but that was not enough. The maths underpinning astrology made it more convincing. Maths was used to generate reassurance where none existed. To do so, it used data that had little to do with the question being asked. Maths was the arcane language used to interpret a future apparently hidden deep within complex data.

THE PRICE OF EVERYTHING

Since the Agricultural Revolution the market has been the place where goods were traded. Although the amounts of commodities increased, markets remained generally much the same. From the 12th century, however, debts were bought and sold. By the 13th, government bonds were on the market. With the rise of the trading companies in the 17th and 18th centuries there came stock markets roughly as we would know them today. The buying and selling of small portions, shares, of a company or its debts. A trade in pieces of paper, in moral obligations and promises, in trust, in future events, not physical goods, rapidly expanded around the world.

With the advent of the share market, companies became obligated to do whatever they could to deliver dividends (the payments made from company profits to shareholders) and increase share value (the actual price of a share in the market). If a business makes a strategic decision, or fails to react robustly to a situation, it could lead to a fall in its value, and potentially the demise of the business and losses for its owners - the shareholders.

Those buying shares came to realise the importance not just of a company's tangible assets as a measure of its value, but also of its decision-making and preparedness as well as its exposure to outside influences. Share prices became a complex combination of asset valuation, general feelings of confidence and predictions of future events. Just as the weather or the demand for a cargo might have affected the financial success of a trading expedition to China in the 1700s, so would the owners' ability to recover from its loss; their other interests, the skill of their senior clerks, even their geographical location. All of these varying and uncertain elements meant a duty of companies to try to predict the future as best they could in order to reassure shareholders.

Shareholders' main concern was for the value of their asset – their potion of the company - and they wanted to be sure it would increase over time until the point at which they chose to sell it. Businesses had to reassure markets, the dictators of share price, that they were building a strong asset, one that would grow in value, that would be around for years to come. That model served the corporate world well enough for many years. Solid, long-term strategies, risk-averse and prudent planning saw many companies grow for years. The value of a business depended largely on good governance and economically sensible, although in some cases not entirely moral, actions. This generally saw the best-run, most innovative businesses grow steadily in value.

Once, someone buying or selling shares would be well advised to know something of the business they were buying into, otherwise all they were doing was recklessly gambling. But as companies became more and more complicated, involved in more areas, influenced by more events, people outsourced the responsibility to experts. Investment bankers and share traders, often specialising in a particular industry or commodity type, would take an investor's money and buy and sell according to what they believed, with the information at hand, would happen to the value of those shares. A class of professional investors, stewards of other people's money, started to run share markets and as a result influence the price of shares.

Developments in technology also changed the nature of share ownership and trading. Communications technology, particularly the telegraph, radically increased the speed at which investors could buy and sell, and also the speed at which important, market-affecting news was broadcast. This meant share prices now moved up and down quickly. A piece of news about, say, a strike in a steel works would be read by traders within minutes of it being called and they could sell their shares in the company, or buy shares in competitors, almost immediately. The news would see a decline in share price as more and more traders sold their holdings. This short-term volatility meant much more frequent, albeit often relatively small changes in share price. Those movements became profitable targets for traders, especially for those handling large sums of money.

Of course, that was true of those seeking short-term gains. Those holding shares for longer would probably hold out unless the news implied some sort of terminal decline. Short-term loses would also see an increase in people buying a share, if they thought it was good value and that a return to growth was not far away.

Confidence can mean surety, trust or faith; it can mean self-assurance, or it can mean secret and private. In financial markets confidence came to be the single biggest factor in a company's value. Two businesses, superficially very similar, can diverge with one failing and one thriving simply because investors, experts in assessing businesses, declare one to be somehow better run or better set up for the future. Often, they are right, but not always. They can be subject to being misled or falsely informed. The assumption is that there are enough people involved in the market that not all of them could be fooled. The majority, as in democracy, would probably turn out to be right.

Key to providing confidence in a business or a country was the economic forecast – economists' predictions of what would happen to a business and the national economy in which it operates. Experts issuing a reassuring assessment, or a doom-laden one, would have significant impact on market values.

By the early 20th century the economists' job was two-fold. They still reflected on political ideologies and theories of trade, currencies and finance. However, they also predicted the movements of stock markets, the growth of industries, and national economic conditions. These forecasts were based on market and company information, but also demographics, currency exchanges, employment rates and so on.

American economist Roger Babson was one of the first truly influential forecasters. He went on to establish a whole company providing stock market forecasts and reporting. He was also one of the first economists to relate the business cycle (broadly the period of time over which the national economy rises, falls, then rises again – the boom and bust pattern) to the physics of motion – specifically Newton's laws of action and reaction. He concluded that the movements of finance, the patterns that could be seen over time, were as subject to the laws of nature as everything else.

Many others came after Babson; economists studying vast amounts of statistical and other relevant information and producing predictions of various sorts. Two problems could not easily be overcome, however. One, that markets and economics were subject to the irrational natures of people, whether that is people buying and selling shares or people spending in shops. The other problem was that the primary market for such forecasts were business and government. So it was business and government that the economists invariably relied upon for their income and who also demanded that their news be positive.

Economists and their forecasts became the chief source of confidence; of reassurance to those investing in stocks and shares. Global financial markets came to underpin all major national economies. These markets came to rely on the movement of numbers, and as communications technology became ubiquitous, these numbers were little more than electrical signals logged by banks and markets. Today it is all digital, and there is almost never anything tangible behind the numbers. If there is, it is a long way down the chain of transactions.

This quasi-automated production of money from money requires a certain degree of belief, without which the system seems quite illusory. Just as early trade demanded trust, a trust founded often in God or culture, so the new trading, the trading of numbers, requires trust. A belief that needs confidence beyond the type offered by religious belief or shared cultural values. It requires something very similar to fact; something that looks a lot like fact. Fortunately, the sheer volume of numbers involved means that mathematics, the foundation of scientific proofs, can be relied up to provide the evidence required. Irrationality, however, is at the root of most economic activity, at least to some degree, and as such the solidity of maths starts to dissolve and with it faith in these new, ephemeral markets.

Before stock markets - and by extension markets in securities and bonds and futures - economics was a simpler place. As a discipline and area of study it ran largely on tangible factors that could be examined in the longer-term. You did not need economic models and predictions of future events to provide confidence in a farm or railway or ship-full of cotton bales.

In highlighting the difficulties of stock markets, the influential 20th century British economist John Maynard Keynes observed that a farmer cannot "remove his capital from the farming business" one day and then "reconsider whether he should return to it later in the week". Before modern financial markets the farmer's, the ship-owner's or the steel producer's belief in their work, or their ability to convince others of its value, had little or no impact on the supply-and-demand economics that priced their goods. As Keynes further stated, "the Stock Exchange revalues many investments every day, and the revaluations give a frequent opportunity to the individual… to revise his commitments." That revaluing and shifting of capital from one place to another created a chaotic system of market pricing. Now, of course, the markets revalue every second and billions are invested or divested across thousands of assets every day.

In the midst of the aftermath of a financial crisis there is often talk of restoring faith or renewing confidence; in a company, a country, or a currency. Faith in a company or a government bond is not just down to economic forecasts, important though they are. There are other elements that come in to play. Like, for example, bad publicity or an exciting new product. It is for this reason that some observers have claimed that actually a poll of informed but not expert members of the public will provide as sound a market investment as many expert traders. That idea, however, raises the important question of what humans think, and why.

12

THE BUSINESS OF THE MIND

"It would be an illusion to suppose that what science cannot give us we can get elsewhere."
– **Sigmund Freud**, *The Future of an Illusion*

Since the Industrial Revolution, progress and growth were, over the medium-to-long term, certain. The Enlightenment, science, and wealth all characterised Western European and American life. That much of this had been at the expense of the other populous continents is in little doubt. Africa had provided much of the raw materials, including human workers, which fuelled the technological innovations. Just as the geopolitical West had largely entered a cycle of prosperity, so large parts of Asia and Africa had seen cycles of poverty as they were over-run by better equipped, better armed nations who took their wealth and gave little back.

As the 20th century approached, many in the West were feeling overwhelmed by the speed of developments. There was so much that was new. Science and philosophy, from Darwin to Nietzsche, Tesla to Curie, had undermined the old orders of God, superstition, and even of human superiority in the world. Within a lifetime electricity, photography, medicine, mass communication, mass transportation had all changed the world dramatically. People were living closer together, they were living longer, they owned more, they were connected by railways, they were more educated and had greater access to information.

Impressive and exciting as all this was to some, there were those who felt it was too much. The world seemed a hectic, commercial place. Old certainties had been chipped away at and people yearned for solace and explanation. There were movements that revived old traditions, from human craft to paganism, but the momentum was with the modern, the industrial. Even religion appeared to endorse industry and productivity. God blessed those that helped themselves; that worked hard and improved themselves so that they might improve the lot of their family, and their community. Hard work, of whatever type, was an admirable thing that brought the worker closer to God.

Workers were also experiencing significant change away from the factory, warehouse or mine. They had leisure time and disposable income, something they had never had before. They watched sport, went to the musical hall or vaudeville, and went on holiday. They also bought products - the necessities, of course, but also extras. They spent on the decorative rather than the merely functional; the tasty rather than the simply nourishing. Even poorer families were slowly becoming consumers.

Private enterprise dominated the economy of the West. Every conceivable item was produced, harvested, transported or otherwise sold by a private company. Private finance had entered every part of life. At the same time, the national economy faced challenges, but chief amongst those challenges was ensuring continual growth. The GDP had to grow in order to keep up with increasing prices and to gradually improve the lives of everyone in society. To achieve that, the underpinning of the economy, commerce, had to grow overall. Given the economy relied ultimately on the consumer, how could they be made to continually consume and consume more? The consumer had to continually earn and spend; they had to be as useful as they could be in the cycle of economic growth.

Adam Smith had proposed that in order to be as effective as possible a business must anticipate its consumers' requirements and satisfy them quicker or better than any competitor. Being able to predict consumers accurately would therefore mean greater success for the business and the wider economy. However, predicting people was notoriously hard, especially since they started living together in cities, going to school and thinking for themselves. Perhaps the solution was to influence them; to create a need to fulfil, rather than just wait for the need to arise naturally.

READING MINDS

As humans formed settlements and moved from nomadic to static populations, in common with all revolutions, the Agricultural Revolution brought unintended consequences. As well as the physical ones such as new and more virulent diseases, there came something different; something that for millennia would remain a mystery. Like all of these pre-historic and ancient mysteries, humans would flounder around looking for explanations in spirits and religion until they had the mental and physical means to truly start to understand it.

Living together changed humans. Humans, uniquely amongst primates, established a co-operative, social method of raising children, rather than exclusively by a parent. As humans evolved, human children required too much care and too many resources for just one person to adequately provide. Living in larger, static groups created a diversity of influences on the early formation of personalities. Slowly, unknowingly, human behaviour changed, adapting to living amongst others and the social orders that started to appear amongst ever-larger populations. People became subject to new influences like peer pressure and social status.

As well as physical weaknesses from disease and changes of diet that the settlement life brought, there were also social fragilities. Someone's place in a group could be influenced by many traits; their inclinations, education, and expectations combined with their physical and inherited characteristics. Previously a tall, strong person might see their life, to some degree, shaped by that. Now their personality would also come into play. They may be physically strong, but displaying qualities of patience or bravery would affect whether they should apply that strength to farming or fighting. Success in their occupation would affect their social status and that too would influence them. Equally, success in one area would mean they ignored others. Displaying the physical and personal traits to make someone a good miller might mean they never bother to pursue other skills or talents.

The settlement would also introduce competition; not just in a commercial sense where one's livelihood might depend on it, but in a social sense. Measuring oneself and one's achievement against those of others in the settlement, or even in neighbouring settlements. Success would influence confidence, which could bring about yet more success, or potentially lead to hubris and humiliation when failure inevitably struck. What would success or failure inspire in someone? Would they be generous or selfish, gracious or mocking? Human strengths and failings were seen as beyond human understanding.

The relationship between body and mind or soul, the nature of good and evil, who a person really is, preoccupied ancient thought around the world. Of course, at the time, the nature of a person, whether they were good or evil, wise or foolish, was invariably a gift (or curse) of the gods in some way. It was more a case of understanding who they were, or predicting who they would be, than changing or influencing it.

Once social conventions had been established, they were hard to break. Families of farmers tended to be farmers for generations, until they had to move off their land or a son went to war. For others, usually higher up the social order, there may be more choice.

To deduce what talents and traits the gods had bestowed on a person, the experts were consulted. From the earliest records there is evidence of priests, shamans, mystics and sages entering trance-like or meditative states in order to commune with the gods. As well as making representations and receiving predictions, they claimed to gain insight into the nature of an individual.

For those believing in the importance of astrology, the alignment of stars and planets when a person was born or came of age would influence their talents. This was vital in order to extrapolate what sort of a personality they were and what roles in life there were suited to. The stars would indicate if someone was wily or empathetic, studious or active, and as such whether they should be a soldier or a scholar, whether two people should marry or undertake a business venture.

In China, India and elsewhere, science (including medicine), religion and philosophy were intertwined for millennia. It was understood that the physical and non-physical were innately linked, and so the study of one necessitated the study of them all. The divine created the conditions for the earthly and it was the duty of special, learned people to understand them.

Chinese medicine has believed since ancient times that the mind (as opposed to the brain) and the body are intrinsically linked. The first physician in China to use anaesthetic, Hao Tuo, recorded how his treatment of a man with scabies was only possible when he replaced the man's fear of the treatment with anger. Fear, Hao Tuo proposed, would prevent the treatment from working, so he annoyed his patient to the point of rage, and then treated him. The treatment itself would have been the same regardless; it was the man's mind, his personality that made the difference.

In Western antiquity, Socrates considered how the mind learned. Aristotle proposed that the mind was built, and that learning was based on education and experience over time rather than something innate from birth. Many theories persisted over time and the human mind and personality were long held to be matters for philosophical debate. Assorted explanations would wax and wane throughout history. Overall, however, the prevailing view continued to believe that human traits were unchangeable, some inherited, and some a matter of superstition.

These ideas of divinely gifted personalities and astrological influence held sway for centuries. For much of that time the brain was not even considered the source of someone's personality, their memories or learning. Science, however, questioned this idea just as it questioned everything else. Like so many other areas of human interest, what was once the concern of philosophy - life, the mind, consciousness - became that of science. In conjunction with physical medicine, ideas of the nervous system and the physical nature of the brain came to correct many long-held misconceptions. David Hume and other Enlightenment thinkers explored ideas of what made humans human. By the late 1800s, centred in particular in America and Germany, science, or an application of scientific method, had started to dominate the study of personality, behaviour and thought.

Benefitting from a middle-class upbringing and a nation starting to invest heavily in education, the German medic and academic Wilhelm Wundt introduced empiricism to the study of the mind. Until his work in the 1870s, psychology had remained a concern of philosophy. At the University of Leipzig, Wundt established the first laboratory designed for psychological experiment. He would go on to produce groundbreaking work on the psychology of language and knowledge.

The first to formalise ideas around perception and understanding, Wundt examined the conscious mind and the relationship between the physical and the mental. He related anthropological notions of culture to Darwin's biological theories. He also proposed that psychology could only attempt to explain the effects of an experience. It would not be able to predict that one would follow from the other.

Wundt was the first person to describe himself as a

psychologist. He started an intellectual race to discover more about what made humans act they way they did, and many important figures followed his work. These would include some of the most famous figures in the fields of psychology and social science. They would define psychology for generations and become feted around the world.

These figures included the Russian Ivan Pavlov, whose work covered impulse and conditioning. He examined how involuntary or subconscious actions could be triggered through repeated association with outside stimuli. He was so important that the Soviet government forgave him his repeated denunciations.

Sigmund Freud, arguably the most famous of all students of psychology, formally established psychoanalysis; the use of psychology as a therapeutic tool to help those struggling with anxieties. He popularised ideas around repression, sexuality, and subconscious feelings of desire, guilt or anger. He notably defined the id (the irrational, random, impulsive mind), ego (the rational, pragmatic mind), and super-ego (the partly unconscious driving, ambitious mind).

Freud's intellectual collaborator-turned-rival/critic Carl Jung furthered the work of the older man with ideas such as the collective unconscious - the existence of unconscious ideals and concepts shared throughout humanity irrespective of culture or geography. He also defined the personality traits of extraversion and introversion, and even the very notion of the self.

Psychology uncovered the motivations and triggers, the interpretations and processes that humans share, consciously and subconsciously, learned and instinctive. It understood there were elements of psychology unique to an individual, but also that there were aspects common to large groups or even to all humans. It gave logical explanations to social, economic and cultural influences, reactions, emotions and behaviours.

An understanding of the mind, whilst slow to gain widespread acceptance, especially given the conflicting theories over certain aspects of interpretation, started to be applied in fields outside academia and health. Anything that dealt with people started to see some use in at least trying to understand that with which it dealt. Including business.

HOW TO MAKE PEOPLE SPEND

Walter Dill Scott was born poor, living on a farm in Illinois until he went to university in the late 1800s. He went to Germany to take his PhD in psychology, returning to Illinois to teach in Chicago. Instead of teaching psychology in the conventional way, however, he applied it to the 'practical' world of business, particularly management and advertising. In 1903 he published *The Psychology of Advertising in Theory and Practice*, the first written analysis of advertising and how it worked (or did not work). Scott was another American pioneer. His work on critical judgement, sentimentality, emotion and persuasion radically changed advertising.

Scott claimed that advertising was more science than art – an idea that remained controversial in the industry for many years. He suggested that practitioners should consider what lies behind an ad, rather than the typeface or imagery that generally preoccupied them. In this, Scott took what had been up to that point a medical and academic concern and applied it to something that most people thought was just a part of everyday life.

To Scott's mind, psychology offered the key to understanding what someone bought and why, how they reacted to what they saw and how receptive to suggestion they were. Understanding this would revolutionise any business that might spend substantial sums trying to promote and sell their wares. Scott asserted that data analysis would reveal why consumers preferred a product or what advertising was more effective. With enough data on enough subjects, a formal picture of what types of advertising had what effect on what groups could be realised.

Informing an audience that your product exists has gone on for as long as there have been markets. A mosaic dating from around 35CE was found in the ruins of Pompeii. It claimed that Umbricius Scauras' fish sauce was of superior quality. Although probably not the first advert, it is the earliest known example. Ever since, the names of producers and manufacturers have been attached to a product in order to imply authenticity and quality.

The format and sophistication of the message advanced over the centuries, but the idea remained much the same. If a consumer that had a need for something was exposed to an appropriate advert, there was assumed to be a good chance they would purchase. Over time, the influence of ruling elites came in to play. They became tastemakers and influencers. What monarchs and aristocracy consumed invariably became the height of fashion. Makers expended a lot of resources courting influential customers. Those that sold to key individuals or families would gain a reputation for quality, style or reliability.

Word of mouth was a powerful force and reputations were hard won and protected. This was effective in small-scale local economies and small, niche markets (such as home furnishings for the wealthy), but as industrialisation enabled mass production and transportation built national and multinational companies, products would be sold to ever larger and more diverse markets. More businesses started to tailor their offerings to sections of the public (known as market segmentation) with different incomes or preferences. Products, ostensibly the same, would have fewer features or be a slightly lower quality material in order to profitably sell them to a more aspirational, middle-class market.

It has been the goal of advertisers ever since to better understand both the target market for a product or service, and how that market will respond to certain messages. Segmentation became a vital part of advertising, trying to find better ways to reach specific groups, and understanding more about who they are. This could be in terms of income and expenditure, but also psychologically so that messages could be tailored to have the desired effect.

Businesses would start to market themselves using different products, advertised in different ways, to different audiences. Those responsible for advertising and selling would study tax and census records to see where customers with certain income levels lived. Advertising then attempted to physically follow the prospective customer, placing billboards or press adverts accordingly – the middle-class clerk reading the newspaper or the working-class housewife in the high street. Repeatedly trying to inform their potential market that their product was available and the best.

Aside from getting consumers to buy their product, businesses faced a further challenge. In order to turn an ongoing profit they had to sell constantly, and ideally sales had to increase over time. It was a concern for many companies, however, that once a customer had bought a product, they may not buy another for a week, a month, or even many years depending on what was being sold. Worse, customers might buy from a competitor next time around, and if they were perfectly happy with that why would they change?

Business and the economy needed continual growth. But consumers had their needs serviced, and so why would they buy anything else? Why replace a perfectly good chair or change the brand of soap they use? Having fulfilled needs, businesses then sought to create needs. Advertising had to move from simply informing a market about a product to selling - to persuading. That brought forward a range of techniques, not all of them ethical, to keep consumers spending.

In 1898 the merger of three baking companies in the US formed the National Biscuit Company; what would later be called Nabisco. Having developed a new form of packaging that kept their product fresher, they then moved to create a new type of cracker, largely by the application of new industrial methods. It was lighter and crisper than any other on the market. These two elements combined – a new, lighter cracker that could stay that way for longer. Up to this point crackers in America were sold in stores by weight, scooped from barrels and placed into anonymous brown paper bags. Nabisco's new cracker with its innovative packaging needed to differentiate itself. Why would people choose their new product over the traditional? The company created a brand. A name and image that set it apart. The brand, Uneeda, saw unprecedented nationwide advertising, and unprecedented sales.

Alongside more conventional advertising, Uneeda played on a familiar, homely image of a boy sent to the store in the rain to buy biscuits (he became known as the Uneeda Biscuit boy). They sought to tell housewives that they could save time having to bake their own biscuits by buying ones that were just as good. There were even Uneeda eating contests. The campaign, created in part by advertising agency NW Ayer, was a huge success and changed the market. It was the first example of modern branding and an early inspiration for using psychology in advertising.

Advertising became an industry in itself, advising businesses on the most effective campaigns, the look and message, the combination of locations and types of advert. Having previously studied customers almost solely on their income, advertising started to see the wisdom in understanding more about those interacting with their work.

Whilst advertising would continue to view income as a main indicator of education, taste and rationality, they embraced psychology as a means of understanding what affect their adverts would have. It used images and messages to provoke emotions as well as to convey information. Adverts would inspire envy, happiness, curiosity, even antipathy. They convinced people that they were unhappy and that buying their product could assuage their misery.

These techniques were applied not just to business but also to politics and public service messages. With the advent of radio and then television advertising they started to ask not just who was consuming that media, but what their state of mind would be at that time.

As well a creating adverts that were original, techniques of persuasion became important. Basic motivations were manipulated. Buy this because it is what famous people use and it will make you a bit more like them; buy this because if you do not, others will mock you. Basic ideas of peer-pressure and social aspiration were used to sell. Marketing and advertising created a need, inspired a sadness that could only be served or solved by the purchasing of their product or service.

Like many minerals, diamonds have no inherent worth. They have few practical uses; being incredibly hard their main application is in cutting other hard substances. They are impractical as a form of currency, and even their most famous asset, their appearance, is largely down to human craft. In the 19th century a group of British diamond mine operators in South Africa created a cartel to maintain the image of the diamond as incredibly rare and therefore of great value. The group, De Beers Consolidated Mining Ltd became one of the world's leading extractors of and traders in diamonds. They controlled supply and created demand. In the teeth of the Great Depression of the 1930s, and falling diamond prices, the company employed NW Ayer to arrest diminishing sales. At the time, diamonds were just one of the precious stones popular in engagement rings (some estimates put it at around 10% before the 1940s), but as fashions changed, so did the type and size of stone used. NW Ayer set upon the tactic of persuading young woman that if a man really loved them, they would buy them a diamond, whilst also telling amorous young men that if they really wanted to display their love for their betrothed, they needed as big a diamond as possible.

The company set about a long-term plan of selling the idea of the diamond as a symbol of profound, ever-lasting love. Ayer set up photo shoots and newspaper articles centred around socialites and Hollywood stars who, upon getting engaged, had ever larger stones on their fingers. They used works by the likes of Dali and Picasso. They even sent lecturers into high schools to espouse the value of the diamond ring to impressionable girls (who would want the ring) and boys (who would spend as much as they could on them).

The diamond ring became the embodiment of aspiration – as a small part of a more glamorous lifestyle available to 'normal' people for that once-in-a-lifetime moment. In 1947 the Ayer copywriter Frances Gerety (in those unenlightened days women in advertising were only given work on projects deemed as being aimed at women) coined the tagline "A Diamond is Forever". Although initially met with indifference the line stuck and Gerety worked on De Beers' ads for the next 25 years. Her work became part of popular culture, the lasting testament to any successful marketing. More than that, Gerety turned the diamond ring, a useless little mineral fragment, into an expression of emotion. Something that brought joy and that symbolised something that was otherwise inexpressible.

The campaign was so successful that for over 60 years the diamond has remained the default choice for the engagement ring. In the 1980s the same agency promoted the (entirely arbitrary) idea that a man should spend the equivalent to two months pay on a ring for his future spouse. Amongst the pantheon of successful marketing campaigns, this is perhaps one of the best and it typifies how advertising sought to manipulate people's emotions in to parting with money for something they did not really need.

Although successes like Gerety's were more down to personal insight and instinct that any academic analysis, it was clear that psychology had a key role to play in making people buy. Psychology would be applied elsewhere in commerce, principally in understanding and motivating workers (with and without their permission). New ideas around management and leadership emerged, with psychological bases of how to organise offices and teams and make them more efficient. It would also be applied in a publishing phenomenon now known as self-help. In America, the land of the entrepreneur where success is just a matter of application, the likes of Dale Carnegie and Napoleon Hill became publishing and public speaking sensations with their ideas around how people perceived themselves and how to achieve personal and professional goals.

Attempts to understand human behaviour, primarily for their own benefit, had evolved into a method of manipulating them. This new discipline could manipulate them for good or for ill. Frequently it manipulated them for profit. Yet economics, the principle study of business, production and money, with its adherence to personal advantage and self-interest, struggled to account for people being manipulated against their own interests. Economics relied on people making an informed choice of what was the best use of their money. Yet psychology suggested that they might be persuaded by a variety of means to just spend on almost anything, whether it served them or not. So how to incorporate this phenomenon? How to account for people buying what they did not need; buying what was frankly of no use? How to account for their irrational behaviour?

13

A CRISIS OF PREDICTION

Economists, like their predicting forebears, know the value of their unique, complex, and hard to interpret work. They protect their discipline. They are, after all, only human and they have bills to pay like everyone else. They are also in a singularly advantageous position, working closely with those in power and well versed in the concepts of supply and demand, to realise the maximum remunerative value of their advice.

Once seen as trusted, sometimes brilliant, authority figures, the public perception of economists has been seriously, perhaps irreparably damaged. To many, they are seen as having failed to protect the public from the punishment of mismanaged financial markets. Markets over which the people themselves had virtually no access and little knowledge. Surely it was the economists' job to raise the alarm? They laid claim to uniquely understanding this aspect of the world. An elite group of global economists have tarnished their entire profession. They played a part in building an ivory tower populated exclusively by politicians and big business and have been reluctant to accept responsibility for the shortcomings of their industry. They have allowed their profession to be seduced by power, and to be abused by those with unscrupulous aims. The language they use bamboozles and excludes even the well-educated. They do not readily share their insights with the wider public and they may never be forgiven for it.

It could, of course, be argued that economists never sought a duty of care in the way politicians have. Nor do their actions directly affect the markets and the economy in the way business does. They must, however, accept their wilful blindness to the limitations of economics that many in their profession suffered. The also knowingly overlooked conflicts of interest in being paid by those they were supposed to analyse and evaluate. Few economists have been able to convincingly refute these charges, and that has only added to the animosity many feel towards them.

Before the Enlightenment, priests were seen as trusted, wise figures with near-absolute authority. Their language - Latin as well as a familiarity with other ancient languages – mystified and awed a public that were by large illiterate. It endowed the priests with the assumption of intelligence - that is not to say some were not intelligent, only that their wore that intelligence quite visibly, and in some cases, exploited it. Despite their divine gifts and knowledge, however, the priests repeatedly failed to protect the people in times of war or famine. They built an ivory tower with politicians and monarchs. They viewed themselves as infallible, but in time they were found out or at least replaced.

Where Latin was the rarefied language that both secured and displayed the status of the priests, so economists have used maths to confuse and convince the world. They have convinced governments, banks, businesses and even themselves that maths holds the answer. That their complex arithmetic models are right, and if they are proved to be flawed, it is because they had not yet collected enough data to make the models truly robust.

One of the oldest mathematical texts in existence, the Chinese *Arithmetical Classic of the Gnomon and Circular Paths*, written around 1000BCE, accurately plots and predicts the movement of stars and planets. So vital was an accurate knowledge of these heavenly activities that around the same time the emperor ordered the construction of a large armillary sphere. The sphere was made of solid bronze - a metal that at the time was the basis of all Chinese currency. It was quite literally made of money.

Mathematics underpinned astrology just as it underpins economics. It used accurate models to predict the movement of stars. The stars move in precise, predictable ways but the relationship between them and the human affairs they were said to influence was utterly unfounded. Just as economics tracks the movement of money and the effects of policy, it cannot assume precision where precision does not exist. Maths requires the rational and exacting. Markets are neither of these because they are prone to the influence of rumour, personal relationships, the weather and the irrational assumptions of humans. Yet neither astrologer, nor priest nor economist would do anything to deliberately undermine their work, their mathematics, or their status. They were employed to advise, and advise they did; it was for wise, judicious rulers to act or not to act accordingly.

Despite the public ire directed towards economists for failing to issue warnings to those in authority, the 2007-08 financial crisis brought a bumper payday for many economists. They were charged with explaining what happened and how it might be prevented in the future. Their work started to enter the public domain, through articles, books, media appearances, even films explaining how banks had almost destroyed the global economy and deprived millions of their incomes and homes. Whilst the banking system was clearly the culprit, governments, central banks, and regulators all had their own economists who had to take some blame.

Leading earners in the aftermath were those economists that could point to having suggested years earlier this sort of thing might happen. Those that had warned that if risk was not properly 'managed' or 'financial instruments' were not understood by those investing in them (or, indeed, by anyone else) something would go terribly wrong. Some of those economists were iconoclasts, others were lifelong skeptics of the free-market, but almost none of them were close enough to the powers that could have prevented a crash that damaged millions.

The mathematics economists use to predict markets was the same mathematics used by banks to measure risk and probability. The use by the banks of even more complex, high-level, computer-driven maths to create financial products no one understood made the problem of economic prediction all but insolvable. Predicting the future value of real estate is hard enough but combine it with the value of lots of other bits of real estate of varying sizes and it becomes effectively impossible. Banks used complex models to predict economic outcomes whilst at the same time creating investment products using similar techniques that would defy prediction.

In the early days of the financial crisis, investment bank

Goldman Sachs claimed that the problems it experienced, and which did huge, almost irreparable damage to the company, were beyond prediction. That what was happened was so unlikely it was like winning the lottery 21 times in a row, over successive days. The only possible conclusion being that what Goldman's thought was a calculable risk, and as such something they could manage, was no such thing. The bank's models worked most of the time, but in the time of crisis, they did not because the usual outcomes suddenly become very different when under pressure. This is the so-called 'black swan' event – an event or situation that you only know can possibly exist when you see or experience it.

The economic models, the predictive formulae of banks and central banks, could not bear the stress of these unforeseen circumstances. They could not cope with the uncertainty. But these models, reliable much of the time, had provided a false sense of security – in business and in government. Added to the fatal mix was the mingling of banking, politics, regulation and economics which saw those working in one area often move into another. The revolving door between finance and politics was particularly evident in the post-war years.

Increases in computing power increased the mathematics physicists could do to uncover the patterns of nature. It also increased the maths economists could do to uncover the deeply hidden patterns in the market; patterns that they were certain were there, if only they could be found. The economics profession now made data acquisition and analysis its core goal. Economists were determined to crack the code despite the lack of evidence that there actually was any pattern to discern. Economics was certain that more data on more economic elements would eventually reveal a truth, a complex pattern with which to predict the future. In a similar way, technical analysts in banks studied the historical performance of an asset, rather than its current value or other factors. Their influence within banks increased alongside the increases in computing power. Banks moved from taking a broad, judgment-based view on investment to an almost complete confidence in technical analysis.

Pythagoras believed that because theoretical numbers were easier to deal with than the messy practicalities of real numbers and values there was something of the divine about them. This belief was the birth of numerology and has been popular, to varying degrees, ever since. People have continued to seek to deduce patterns in all sorts of numbers, sometimes by assigning values to religious and secular scriptures, and by applying complex formulae.

In the Darren Aronofsky film π, Max, a number theorist, believes everything in the world can be explained through numbers. During his work he stumbles upon a 216-digit number (216 itself being special in maths - a so-called untouchable number) which seems to be the key to predicting, amongst other things, the movements of the stock market. Max is also told that some believe the Torah to be one long divine numerical code and further analysis shows the holy book contains the same 216-digit number. Capitalists and believers, with equally extreme fervour seek out Max and his number. The latter group declare it to be the name of god and a precursor to the coming of the messiah, whilst the former need to own or destroy it for their own profit. The whole mess is only resolved by destroying all record of the number and some extreme surgery.

The desire to believe in patterns in the world, patterns that repeat and so can be used to predict, has made many in power blind. In some sectors of society that desire has become an essential need; an article of faith that these patterns simply must exist.

In the public imagination, the 2007-08 crash was a collective failure of an elite group to spot what was clearly going to lead to a crisis and prevent it from happening by any means necessary. That elite included economists who were as culpable as the bankers who chased profits at high, incomprehensible risk and the politicians who deregulated allowing them to gamble with public money. Economists, however, do not earn their living from the public. They are invested in by our rulers, and it is common for all humans, ruler or ruled, to fall victim to the sunk cost fallacy – the inclination to chase your losses, that the more you invest in something, the harder it is to abandon it.

Even those economists that did see something wrong with the financial sector in the run-up to the crisis did not realise the scale of what was to come. Mainly because banks were, and remain, secretive about how much money they have invested in a particular market or sector – in their terms, their exposure to something. Banks consider this a commercial secret, despite the very public effects it can have. They keep this secret just in case it does go wrong and that self-fulfilling spiral of lost confidence destroys their investment completely. This ensures that anything even approaching accurate prediction about the effects of a particular asset losing value is all but impossible, even if economic models are accurate and robust. Even if, somehow, some economists really did know how much all of the banks had collectively invested in their array of flawed products, and if that economist had gone public, they would probably have been dismissed by those with the power to do something about it. Written off as freaks, contrarians, or as having some ulterior motive for talking down confidence.

No big business or government is willing to operate on a hunch or notion anymore. Both rulers and public, consumers, leaders and voters no longer trust rule by wisdom. There must be proof. Evidence, invariably backed by robust mathematical models, that the advice they give or take will turn out to be right, or will at least come with a plausible excuse if it goes wrong. Paul Krugman, the Nobel Prize-winning economist, accused the economics profession of "mistaking beauty for truth" (that is the beauty of a well-worked, intricate, elegant mathematical formula).

Economics and those in its thrall have accepted the uncritical assumption that numbers and patterns alone hold the answers. Former World Bank Chief Economist Paul Romer said that whilst maths can help clarify economic thinking, it is not the only determinant. Furthermore, the high level of mathematics, a relatively new aspect of economics because of its reliance on computers, also excludes others wishing to contribute valid ideas. Those outside economics are excluded if economists will only ever consider something that can be expressed in complex formulae and arcane terms. All of this gives economics an authority that appears unquestionable. It prevents outsiders from challenging it and so breeds complacency from those within it. It has the outward appearance of a divinely gifted authority and well-defended status akin to the priests of old.

Economist Robert H Nelson, in his books *The New Holy Wars* and *Economics as Religion*, has examined the similarities between the origin stories, prophets, prophecies and laws of religion and the world of economics. He examines how, as the acquisition of material wealth has supplanted the blessing and approval of God as central to national and individual concern, so economics has replaced religion. He sees parallels in how religion developed to explain the world, and economics has fulfilled the same, modern role. He considers divisions like Protestantism and Catholicism alongside capitalism and Marxism. Whilst sometimes a conclusion or assumption too far, Nelson's work clearly identifies a human need for these guiding, predictive, mysterious rules for living.

Others would point out that religion has been replaced less by economics and more by science and it is in that regard that economics seeks to emulate the latter. Religion is now largely a personal rather than a national matter, certainly in the West. Whilst many politicians in even predominantly secular countries are reluctant to say they have no religious belief, few would be so bold as to say their only response was to pray extra hard for a solution to the nation's problems. Generally they need a hard, fact-based plan. Something logical, akin to science, but given that science struggles to predict the irrational, economics fills the demand. Economics has made the journey from faith to fact, philosophy to science. Economists have co-opted empirical scientific terms and behaviours.

Mathematical models, whilst superficially attractive, do not necessarily equate to scientific standards of proof on their own. In science, theories and models can take decades of work in order to find evidence to prove or disprove by experiment; something that is not readily open to economics. Theoretical physics needs experimental physics to add substance to their models and hypotheses, but experiment has practical limitations. For example, finding evidence to suggest that Einstein's prediction of the existence of gravitational waves was correct required complex technical equipment, as well as the appearance of a ripple to measure, that were not available until over 60 years after Einstein's death. Without this sort of standard of proof a complex model will go in search of evidence, selectively, anywhere it can find it.

As well as an unquestioning over-reliance on the pseudo-scientific nature of the work of economics, the 2007-08 crisis also highlighted the long-standing lack of scrutiny into who was delivering economic forecasts. What was their agenda? Who was paying them? Was it right to have someone both advise and examine a bank? How close were those that regulated, that were charged with protecting the public, to those that were making high-risk profits?

Does all this mean economics is, like religion, something of a dead-end? Destined to still exist for decades to come, a belief to many, but increasingly irrelevant in decision-making? Or, like astrology, that it is the forerunner, an antecedent, to something more reliable? Perhaps economists, or more importantly, those that rely on them, will accept the limits of the discipline rather than insisting on its potential accuracy. That economics provides an insight, but not necessarily the full facts. That after generations of seeking order, of searching for patterns, there is little compelling evidence for either in many important areas.

In attempting to mimic physics in its use of maths and models, economics has frequently failed to acknowledge when it got its predictions wrong. It further failed to develop new practical applications rather than high-minded theories. It is perhaps in this area that it most revealingly displays its divergence from science.

It may be that seminal economist Alfred Marshall (who would no doubt be disappointed at the way maths has been used to exclude non-economists from understanding economic issues) simply got it wrong in trying to make economics too much like a science on a par with physics. The evidence that humans in groups and in markets behave like steady physical systems drawn towards an equilibrium may have been compelling, but so was the evidence that the position of stars influenced a person's character or good fortune two millennia ago. No doubt many people were reluctant to let go of those reassurances as well.

Economics may not claim to be all knowing in the way religion has, but nor does it willingly accept its fallibility in the way science does. In physics, believing something to be true will never make it so. In economics, markets have often been altered, even established, by a common belief in the value of something.

Equilibrium is a term commonly seen in economics. In physics it underpins everything related to forces and motion. That is nature. Not seeking equilibrium is effectively impossible. That an object working against a force, say gravity, seeks a stable state of rest, drives everything. Economics is not nature- it is human-made. Money is a myth we invented to facilitate a more comfortable, convenient life. There is no reason money should obey the same rules, follow the same predictable patterns of behaviour. The rules of physics, chemistry and biology cannot be skewed, managed, biased or assisted. They just are, and we can only observe, measure and attempt to understand. Economics is quite the opposite.

It is the belief in economics as a true science that saw economists, in collusion with politicians, impose Western-style free markets and practices on the former Soviet bloc in the early 1990s. Those countries, isolated by Communism for over 40 years, with little experience of capitalism, were simply not ready for it, and in some cases they may never have been. The economists and the politicians cover for each other to this day; economists saying it was political failure, politicians saying economists used the wrong models. It amounted to the same result: countries forced to work the way the West had, to varying degrees of success, for the last 50 years and which are still, 30 years later, struggling with nepotistic, authoritarian governments and grossly unequal, corrupt economies.

Science reaches its conclusion after study, debate, vigorous challenges and experiment. Many scientists spend years working on a theory that is ultimately disproved or, at best, a dead end. The economic experiment of post-Soviet Europe and Eurasia appears to have yielded a few lessons, little regret and almost no useful change. It is hard to say it has improved the lot of the majority of citizens to the level it promised.

The determination of economics to emulate science by finding patterns in vast quantities of data in order to develop more accurate predictive models necessitates a big picture, distant, detached view. So it is no wonder economists have a reputation for not understanding the small, real-world, everyday indications of what might happen whether that is people unable to pay their mortgages or oligarchs buying up state assets.

Again, it is not always the fault of economists'. They know that often their patterns only work in certain circumstances or within certain probabilities. Like physics, the very small and the very large do not work in the same way. The same theories and models do not work – although they were once thought to. The non-economists running countries and businesses, however, do not always appreciate that, and nor do the public who see economists as part of the problem - especially when politicians and business leaders use them as scapegoats.

In the aftermath of the 2007-08 crisis, economists argued about the conditions that would prevail from thereon in. One standard economic model predicted that during such an economic depression large increases in the money supply (known as quantitative easing or QE) would not lead to inflation. Some economists said this was patently incorrect and more money entering the system would always lead to inflation. Others said it would not. It was the latter group that would be proved, ultimately, to be correct, at least in the circumstances of the time. QE became policy for many Western central banks for many years after the crisis. Science would not have to have that sort of argument as the conditions for science do not change; just our understanding of them. Economics is not a science where everyone agrees on what constitutes empirical evidence or fact. Its predictions are rarely agreed upon; politics, ideologies seize on those predictions that serve their individual purposes and are promoted accordingly.

In the absence of hard facts economics views mathematics as lending it the credibility and certainty it would otherwise lack were it purely a question of philosophy, ethics or politics. Like politics, economics starts with a worldview - a hypothesis of the way someone wants the world to be - and then works backwards. It selects the data and examples that prove the theory right. Were it otherwise there would be no right/left arguments on everything from welfare to bank bailouts to education.

Maths does not vary depending on your point of view, country of origin, or socio-economic background. In politics, and frequently in business, each idea finds its own portfolio of economic evidence and then shouts louder and louder about it. Political ideologies and management strategies are foremost about dealing with people, but they apply economics to work towards an end goal – power or profit. Economics finds itself biased, whatever its original intentions, endlessly tweaking its mathematical models to prove that a given worldview is right.

The degree to which the state or government is involved in planning or affecting economic activity is philosophical rather than economic. The strategic direction of a company is usually down to a few key individuals who have particular views on prudence and risk. Economists of different persuasions are cited as the evidence and their models prove every political doctrine, opinion, policy, decision and trend. They justify all types of commercial strategy from what to invest in or to acquire or whether to downsize or outsource.

It is also in real world applications where economic forecasts are seen for what they are; the tools of decision-makers who are often unsure but cannot admit to it. The problem of economic forecasting is not always within itself but within those relying upon it. Those relying upon it are also not at fault because of their circumstances. Economists invariably publish their advice with estimates of probability, but a clear decision is needed. Voters and shareholders need decisiveness and have little appetite for nuance or equivocation. So it is that once again, economists blame the simplification of politicians whilst politicians blame the inability of economics to predict or give a simple answer to a simple question.

Strategy and policy in big business and national government can often take years to implement. So the only way to effectively reassure those reliant upon them is to forecast what will be happening in two, three, five, ten years, and set their course accordingly. In this there is also an element of self-fulfilling prophecy. If a car manufacturer is told that people will be buying crossover vehicles, companies will design and prepare to ramp up production accordingly, and in doing so, competitors will follow and effectively force the market into buying them.

Science is the exploration of the natural world; unbiased in its observation of the fundamental elements of the physical. Economics and every aspect of it are human constructs; no more natural than literature, cities or sport. What started out as a philosophical investigation into humans, money and trade fell in love with science and maths. Economics is at least as much about society, ethics, and politics than data and complex formulae. The flow of money is not like the flow of water under the effects of gravity or electricity under the influence of potential differences because these are not human constructs. Water, gravity and electrons are parts of the physical world that do not think for themselves. Money is an invention of unpredictable humans, but rather than acknowledge the limits of its mathematical models, economics persists and pursues them with ever more vigour.

The people running companies and countries understood that they were dealing with humans; unreliable, unpredictable, strange humans. Economics put psychology, whether in advertising or in people management, under the category of 'innovation' and moved on. It viewed psychology as just a new way to make more money or improve efficiency just the same as the invention of a faster or more reliable manufacturing process or machine. In truth, this revealed the limits to conventional, classical economics.

Adam Smith and his forebears were philosophers; they

understood the human aspect of their work - even if they would later be proved to have misjudged it. Yet economics came to be in thrall to science. It rigidly, pig-headedly pursued a scientific, empirical method. Whilst philosophy may have cause to call upon maths or maths-like formulae in demonstrations of logic, it maintains a human element. Its vast span of interests and influences mean science can only form part of what philosophy does. But modern economics sought to smash such thinking and pummel the world into a neat, rational, predictable order.

Economics and its covetous relationship with science has infected ever more aspects of life, nowhere more so than in government. Over time leaders came to believe that numbers were the only measure of success or failure necessary. Perhaps because business tended to be measured in a few numbers – profit, loss, staffing, assets – leaders tend to believe everything will fall to the same methodology. A military, a health or education system, crime, happiness and wellbeing must all surely be reduced to numbers in order for the success or failure of a policy to be understood. Numbers, with their solid, processable certainty meant they became the only way to assess anything.

Numbers are simple and efficient; they go up or down and they fit easily into headlines that the public understand. Consequently, government became a matter of managing (and massaging) numbers in order to appear effective and retain power. However, arrest rates do not measure crime prevention or social cohesion. Patient waiting lists do not measure health education or healthy lifestyles. Exam passes do not measure student improvement, mental health or their acquisition of soft skills.

The hegemony of numbers amongst political and business

classes has also seen the equating of numbers to intelligence. An instant recall of the key numbers – exchange rates, employment figures, GDP, approval ratings – demonstrates both a superior intellect and knowledge of who is winning the political or commercial race. The numbers are simple, unarguable, and in an era of rapid change, they offer a clear insight at that moment.

Numbers are indeed quick and easy to understand, but excess focus on them has been at the cost of the factors that just as important, but harder, more time-consuming and more expensive to measure. Everyone knows the problem, but no one can break away. The numbers are just too seductive. The public, at once in the thrall of falling crime rates and rising exam passes, also know that statistics can be manipulated or quoted out of context rendering them meaningless. But without any other information, what do they have to go on? Governments are trapped into short-term policies to improve numbers, ideally for little or no cost, desperate to find efficient ways to change things where generations of predecessors have failed.

It seems strange to us now, but for two hundred years, until as recently as the 1980s, economics was solely run on the assumption that people were entirely rational. They made decisions logically and with all the relevant, available information to hand. Their irrational behaviour was absorbed into a margin of error; small enough to not significantly effect the overall trend – the direction of the majority.

Perhaps it is an indictment on society that it was never deemed necessary for economists to seriously consider the irrational. That actually, despite our protests that we are generous, empathetic, creative, unique individuals, when it comes to money, we are cold, calculating, self-interested and generally predictable. The amount we spend, invest or give away irrationally is so small as to be irrelevant. The evidence, however, suggests that it may not always be so. Even if our irrational sides were marginal, as with Chaos Theory they can have massive repercussions in the right conditions. Is it possible to know when those conditions might arise? Could, by understanding the irrational human, the shortcomings of economics be addressed?

Human thinking and decision-making can be irrational in all sorts of circumstances. It is, however, the long-term, big picture and large scale that gives us the most difficulty. It is why we have come to entrust experts, leaders, organisers, and governments to protect us, plan and make the best decisions for us. They, more than anyone, need an insight into the irrational behaviour of humans, and whether it could be contained.

14

THE UNDERMINING OF ECONOMICS

Economics is important. It provides insights and understanding, analyses and explanations. It is, however, the study of the past - of past events, activities, interactions, causes and effects. Its forecasts are based on analyses of history. As long as history repeats itself, as long as the world stays roughly the same, as long as people are consistent, the forecasts will be reliable.

In times long gone, astrologers applied mathematics to the movement of stars and planets in order to predict human affairs. The most brilliant mathematicians were sought after by the most powerful rulers in order to offer superior astrological advice. In modern times, economists apply maths to past economic, and even non-economic behaviours in order to predict future ones. Once again, the most brilliant mathematicians are being employed, not in research or engineering, but in the business of prediction.

The hegemony of economics is what is damaging; not the discipline itself. Economics' role in growth, and national dependency on that growth, has given economics incredible influence over rulers, and by extension, over us. The influence and importance of economics has made it into a billion-dollar industry in itself, but its work affects trillions of dollars. Governments and businesses pay huge sums to those that can read the economic signs, decipher the maths and interpret the models. Their advice, directly or indirectly, affects almost every aspect of modern life.

In many ways the problem is not of economists' making. They are under pressure from politics and markets to deliver certainty, despite generally knowing they cannot. So, whilst their forecasts are usually bounded by provisos and probability, they are reduced to simplistic, binary statements. Politicians, especially in fast-moving, media-saturated times, believe the public need decisiveness, clarity and action. Markets are weaker when they have to rely on ambiguous information. So economists are at best misrepresented as, or at worst pressurised into, making apparently certain predictions.

Wielding such influence, bearing the weight of so much responsibility means economics must look serious, credible, and solid. And there is nothing more serious and credible than complex science. So economics has pursued the certainty that comes with rigorous, clinical science. It has looked towards physics, and away from its origins in philosophy, as a template.

Physics studies the workings of the natural world. Economics studies the workings of the financial world. Physics sees the interplay of atoms, the flow, movement and effects of forces, energy and work. Economics observes the interplay of industries, the movement and effects of goods, money and labour. Physics seeks to explain the very big (astrophysics and cosmology, gravity) and the very small (atomic and subatomic, quantum). It deals with fundamentals, like the four forces – gravity, electromagnetic, strong and weak nuclear, that explain, and can be harnessed to assist our daily lives.

Mathematical models underpin physics, describing the relationships between inputs and outputs, causes and effects. Mathematical models underpin economics, describing the relationships between financial and policy inputs and outputs, causes and effects. Economics, however, lacks much of the empirical evidence, the rigorous, repeated examination of physics. Physics is strong because it is constantly tested. Physics knows that there is a gulf between explaining the big and the small. It knows that it cannot explain everything. Economics is reluctant to accept such ignorance. It struggles to say it cannot explain everything, or that it might be wrong, or that a better, more reliable model might turn up someday soon. To do so would be to diminish its role, and to diminish the vital confidence it supplies.

Chaos theory is one of the more famous, but also profoundly complicated mathematical theories. Its most famous incarnation, due to its elegance as well as various misinterpretations, is the butterfly effect – when a very small event can have massive, unforeseeable repercussions via a complex chain reaction. The cause and ultimate effect in these events can only be seen in retrospect; never at the time. Weather, predictable only within certain parameters, remains unpredictable in many respects because of the chaos effect. The dependence of weather on the effects of feedback means that a small temperature change in one place can multiply in intensity and have devastating effects thousands of miles away. Or it can peter out to nothing more than a light breeze.

Weather, in the 21st century as much as in the 17th, can have huge effects on the economy. Not just directly on harvests, shipping or oil extraction, but on more complex, nuanced areas like migration patterns and urbanisation. Unlike the cost of oil increasing, which has generally fairly predictable effects, changes in migration are much harder to predict given the vast array of reactions it can have from wage inflation and welfare to social, media and political responses. This is where chaos enters the system. If devastating weather events cannot always be accurately predicted, and certainly they can almost never be predicted a long way ahead, then what chance the knock-on effects? When irrational humans are involved, with their varied, strange reactions to events; with the influence of democratic majorities or vocal minorities, it makes it all but impossible to find an applicable pattern. Looking at what led to what only makes sense, it only appears inevitable, when viewed retrospectively. Piecing together the events and reaction many years afterwards.

Chaotic systems, where random events mean that the finishing point cannot be predicted simply by the nature of the starting point, can be deterministic or non-deterministic. In deterministic systems, where no outside influences affect it, small variations in the initial stage can ultimately bring about very different results, and which way the system will go is unpredictable. Weather systems – as opposed to climate that can undergo outside influences – act like this. Conversely, other systems are chaotic because they are subject to outside influences. These influences can themselves can be random in their nature, and they can bring about unpredictable outcomes from an otherwise predictable system.

One such outside influence can be the very act of prediction itself. This is what happens in financial markets. A prediction can become at once self-fulfilling and self-defeating and it is impossible to accurately know what the effect will be. For example, if someone is seen as a reliable, knowledgeable source, and they predict that the stock market will rise over the coming days, chances are it will rise quicker and by a greater amount as confidence is gained, because of the prediction. That means that the prediction of a rising market ended up being true, but no one can ever know whether it was going to rise anyway, or whether the confidence supplied by the expert inspired it. At the same time, whilst the prediction was correct in its nature, the degree of increase and the timeframe for it could have been wrong, as confidence flooded into the system. That inaccuracy could, in modern markets, have cost millions, which in turn might damage the expert's reputation, leading to a different reaction to their next prediction.

Science develops models that mimic the natural world as accurately as possible based on the current understanding and available information. These models are adequate for our practical purposes, right here, right now on Earth (although sometimes they are required to work beyond that). Fundamental to science is the fact that every effect has a cause, as described by Rene Descartes in the 17th century and expanded by Newton a little later. If an object moves or changes state something caused that observable change and physics aims to describe what the outcome will be. If a body of a certain mass and construction is placed under a force or is heated, what will become of it? The influences of friction, inertia, energy and thermodynamics all create models to explain what happens. These conclusions led to the field of determinism in which, given a finite number of elements, forces and so on, a physical reaction can be predicted. If a ball of a given size is thrown with a measurable force, at a known angle and under the influence of known air resistance, then its final landing point can be accurately predicted.

As has happened many times before and since, science influences philosophy, and determinism in physics was mirrored in theories of human behaviour. Philosophical determinism held that humans were subject to a finite range of influences and as such their behaviour was not of their own determining. Just as science implied a rational natural world where every cause could, in theory, be measured and so every effect could be predicted, so too with people. If you just knew all of the potential influences; if you could measure and model them and the long and short-term reactions, you would be able to predict human behaviour perfectly, in the same way as physics predicts the movement of a ball.

In this assertion, determinism was at odds with the concept of free will; that humans made their own decisions freely, unimpeded and unpredictable by anyone but God. Theology had long wrestled with the notion of free will - the idea that God had imbued humans with free will in order that they might be able to choose between good and bad, right and wrong. Unfortunately, that in turn contradicted God's divine wisdom and goodness in creating a person that would choose to act in the wrong way. Regardless of the explanation, religion could not predict or fully explain the choices a person made. Whatever guidance a religion offered a person, they could, and did go against that advice given the right circumstances. Religion, like economics, is a human attempt at explaining and ordering the world so it is only to be expected that both would fail to predict or fully govern human behaviour.

Science fuelled capitalism, so it is no surprise economics felt it was the new science that measured capitalism. Science had largely supplanted religion in explaining the world, so it is clear why it would be tempting to see economics and physics as effectively synonymous. Physics, after all, owned many of its origins to early astrology, which led to astronomy and the maths that described it. If something so robust, definitive and reliable can come from something so abstract, why should not economics leave its philosophical roots behind and become a science as well? A science of human work and finance. The problem is that science, true science has existed before humans and will continue after, unlike capitalism, language, social class and any other human construct.

That has not prevented some from being convinced that economics, trade, tax, labour, migration and so on are all natural parts of life and the world; as natural as atoms or gravity. As such, parallels are sought with complex economic models derived from quantum physics or thermodynamics. This thinking makes economics concrete; not a collection of competing theories but a real-world explanation of naturally occurring phenomena. So it was that economics pursued this route of being a science. It measured more factors, collated more data, and developed formulae of ever-greater complexity in order to create models from patterns in this data. Models that would predict future economic trends and events. Models that would enable planning, supply confidence, and run human affairs much more effectively.

Richard Freeman, Professor of Economics at Harvard University wrote an article, *It's Better Being an Economist (But Don't Tell Anyone)*. In it he proposed that "physics and mathematics, which attract students who are, by conventional measures, smarter than economists and where the base of knowledge is better established" are at a disadvantage when it comes to the job market. He goes on to say that "economists earn more and have better career prospects than physicists or mathematicians".

Part of the explanation for this is that economists tend to get 'real' jobs with 'real' companies (which have real influence), as opposed to those in 'real' sciences who tend to stay in either academia or other publicly funded areas. Even the academic economist can expect to be called upon by the well-funded sectors of accountancy and consultancy firms and financial services, boosting both their income and that of their institutions. Plus there is a lot to be said for being asked to attend meetings with important people in the corner offices of grand, imposing office blocks.

It is a commentary on the nature and importance of economics in the world that it remains so well rewarded despite the far more tangible, laudable and durable achievements of physics and mathematics. So it is reasonable to assume that some students who are inclined towards physics and maths at school may find themselves attracted to the career prospects of economics. In doing so they bring their inclination, their methodology and skills to a less scientific discipline. As such it should be no surprise that a great deal of economics and finance is, like physics, all but incomprehensible to non-technical people - including politicians and business leaders. By extension, it is has become all but impossible for non-technical people to rigorously question how economics works and be engaged in any productive debate on the subject.

Irving Fisher, Professor of Economics at Yale was a hugely respected economist, writer and pioneer on debt, capital and interest rates. Lauded by his contemporaries he was, by some measure, the best-known economist of his time. His reputation, however, was all but shattered with public pronouncements along the lines of "stocks have reached what looks like a permanently high plateau" and "values in most instances were not inflated". These remarks, little more than ill-advised at any other time, were made in October 1929. The end of that month saw the Wall Street Crash, the greatest drop in share prices in US history, which precipitated the Great Depression. There followed over a decade of global recession, a contributory factor to German economic and social instability, the rise of Nazism, and the start of the most devastating conflict the modern world has seen.

Perhaps unfairly, this example of a respected economist not just failing to foresee the Crash, but actively suggesting it was all but impossible, has been used to criticise the discipline ever since. All the signs were there, observers have since pointed out. As it was, no serious economist foresaw the Crash. Few of them had raised the alarm over what could happen in America (and subsequently the world) as urban populations grew, agricultural industries failed, and market speculation ran wild, as it had throughout the 1920s.

Criticism from within has remained a part of economics ever since. More recent remarks like "[markets have] forecast nine of the last five recessions" (from Paul Samuelson – Nobel Prize winner and MIT economist) and "the record of failure to predict recessions is virtually unblemished" (from Prakash Loungani – former IMF and Federal Reserve analyst) are also used to undermine economics as a predictive profession. So are the failings of the key economists of their time evidence enough to write off the whole profession – as distinct from the discipline itself?

Large-scale financial events, global recessions and

market crashes are only one type of very particular event that an economist might be expected to predict. Judging the profession's overall record must call for a wider examination. Does their inability to foresee a huge economic disruption mean they also routinely fail to foresee smaller ones? Does it undermine their qualification to make macroeconomic plans or business sector forecasts? Is it like a seismologist who fails to warn of an earthquake or meteorologist who misses a hurricane?

Economists typically work on the same reasoning as most of us – if something were true yesterday and the day before, it looks like it is true today and so it will probably be true tomorrow. As a rule, they are right more often than they are wrong, and that is because the world changes only very slowly. A recession, by its nature, is a sudden, erratic shift in the normal order of things. Failing to predict such remarkable events is surely beyond what could reasonably be expected. Human affairs being unpredictable and recessions being chaotic in their nature, why should economists be in a position to warn us? Is it just a case of too much faith being invested in them; faith they did not seek. Are they victims of a wider human desire for certainty? There is, however, another factor: the pressure on economists not to predict economically damaging events.

There is little appetite for doom and gloom in the world of finance. Businesses and markets thrive on optimism, on confidence. There is plenty of proof for this and as Keynes said "Worldly wisdom teaches that it is better for a reputation to fail conventionally than to succeed unconventionally". In other words, predict something terrible, and no one wants to know (like the Cassandra of Greek myth); fail to predict it, and no one will thank you anyway. Markets fear the self-fulfilling prophecy – say a business will fail, and it quite possibly will. So it is probably best to keep quiet.

Even though Jeremiah of Biblical times was in the minority in his doom-forecasting, those in charge still resented the effects his prophecies were having. Perhaps that might also be the case with modern economics. Many (but by no means all) economists have failed to predict serious crashes and recessions, despite it appearing to be one of their most important roles. However, it could equally be the unwillingness of rulers to want to hear doom-laden economic data and advice.

Perhaps because to act on such advice too soon would have serious implications - although not as serious as the implications of the recession - they just do not want to take the risk. Any economic prediction will come with a measure of probability, and rulers tend to gamble on the side that will maintain the status quo; they, like all humans, think that, on balance, it will be all right. If a government or central bank acted in such as a way as to try to mitigate the effects of an on-coming recession it would suggest to the markets that those important, influential parties were worried about a recession. Markets tend to act much quicker and more nervously than governments and their reaction would inevitably lead to more panic and ultimately the recession that was predicted. A self-fulfilling prophecy. However, if the government had held its nerve, if they had taken the chance, perhaps the recession would not have happened, or it would have been less severe.

In the 1960s, psychologist Peter Watson coined the term Confirmation Bias. It referred to a person's inability to rationally assess a piece of information that supports their belief. The phenomenon, though, was not new: the Greek historian

Thucydides noticed it in times of war around 400BCE, as did Dante in the *Divine Comedy*. The father of rational, empirical method Francis Bacon noted that: "The human understanding when it has once adopted an opinion ... draws all things else to support and agree with it."

Watson experimentally demonstrated the human tendency to seek out, and exaggerate the importance of, those facts that confirm what we believe, or want to be true. The economist that finds the evidence, patterns and trends that support business or government goals or existing courses of action will be embraced by power. So important is the economy, particularly the stock markets, that many economists are paid by companies and governments, making their objectivity questionable, or at least worthy of serious scrutiny.

There are conflicting ideas about whether recessions are inevitable. Whether national and international economic policy is just a case of governments and banks staving them off for as long as possible. Of generating as much income as they can in the good times before the storm hits, then hunkering down until the worst of it passes. Like living in the path of a hurricane, but the causes of a hurricane, although not always predictable in the longer-term, are at least agreed upon. Certain conditions have to be present for one to form, and once formed, it will travel long a likely path. But economics as a whole cannot agree on the causes of crashes and depressions because they are not purely scientific and mathematical in nature. Their nature changes as human society changes. Although there are many parallels, the Wall Street Crash was not the same as the dot.com crash or the 2007-08 crisis.

Arguments about the fundamental nature of economics aside, has the economic profession actually had its prediction-making credentials scrapped? Did the collective failure to foresee recessions – events that caused misery to millions of ordinary, low and middle-income people – mean the profession is ultimately of little use? That they had failed in their duty to those that paid them and that gave them power and responsibility?

Blame seems to be shared around depending on who is asked, and it is the circle of finger pointing that has gone towards damaging public confidence in those in charge. Economists have, for well over a century, had the collective ears of big decision-makers. Economists say their assessments always contain a measure of probability and it is for the rulers to decide on an appropriate, cautious course of action in accordance with the chances of events going one way or the other. Government and business say the economists failed to warn them in time or with adequate insistence. Were these instances failures of prediction, of action, or of persuasion?

Both parties, economists and rulers, were responsible for not being alert to the shortcomings of their particular situations. Leaders were over-reliant on a narrow group of advisors and were driven by short-term goals. Economists had come to slavishly believe in their own data and mathematical processes and were seduced by power. Both groups were certainly too insular, unwilling to hear uncomfortable ideas from outside their immediate experience, too protective of their own privileges.

The root of the prediction problem is twofold: the timeframe and the presentation. Any prediction needs to be presented in language the audience can understand, without diminishing key details. The most important detail being how robust that prediction is. What conditions need to be present for it to be true? What are the real chances of it happening? If the prediction does not work out, rather than apportioning blame or seeking excuses, explain why and what can be learned.

Economic forecasts are principally the concern of government and business. Economies, like the weather, are complex, chaotic and only broadly predictable in the very short term. However, the nature of business and government, the time it takes to implement strategy or make decisions, means they need to plan long term. Something put into place now could take years to generate actual actions. The further out the economist is required to forecast, the weaker their prediction, or the broader the boundaries of probability. The further from certainty the advice is, the less useful it is to decision-makers.

Political scientist Philip Tetlock ran a 20-year test on forecasting as part of his Good Judgement Project. He took over 28,000 predictions covering economics, politics, culture and more from a wide range of academics, government officials, writers and others. Tetlock developed a system of contests to find out which predications worked out. He found that, generally, the forecasts made by experts over the course of years were little better than an average of random, chance selections. This performance declined as the range of the forecast moved towards three-to-five years. Almost all government and large business decisions and strategies need at least three years to enact.

Some of this delay is structural, as in companies that place too much hierarchy or bureaucracy into the decision-making process. Some, especially in government and the public sector, is unavoidable, involving huge numbers of people and catering as equally as possible to vastly differing groups. Yet shareholders, employees and voters all demand visible action in the short-term.

This contradiction in timeframes, along with ever more scrutiny of leaders, in turn puts pressure on those that advise them. Leaders need to sound certain in their decisions and about the future so advisors are pushed towards certainty in their predictions. This is true generally, but nowhere more so than in economics. The wellbeing of markets, the country and each individual has become dependent on economic forecasts. The confidence those groups have to spend, to invest, to employ, to grow depends on confidence supplied from government, central banks and others nominally in control.

Assessments and predictions of future economic trends and events used to be largely a matter of judgement; the collective wisdom of a few, select experts and policy-makers. The rise of computing saw more and more complex models arise. The appearance of something akin to science, which deals much more in certainties and facts, started to overpower the human judgement element. Those in charge, aware that their jobs and reputations were on the line, looked for certainty, or at least a plausible excuse.

Private and public sectors generate scores of economic forecasts over the course of a year. Many of them differing in some detail or another. But government tends to have to reduce all of this to just one or two key treasury forecasts. They need to deliver a confident, sure plan. This reduction and the inherent unequivocal statements it leads to naturally produces errors. Statements are misleading, decisions are poor, and the public are further disenfranchised.

Arguably the most damaging result, particularly of failures around the 2007-08 crisis, was a public now too skeptical of economists and other experts they lumped in with them.

Those that pour over statistics and probabilities and use their specialist knowledge to aid big decision-making. Those that deliver advice in complex language bounded by probabilities and never entirely certain or unequivocal.

Whether economists are certain about their predictions or not, those who interpret their advice, who stake a nation's prosperity on it, need to accept the unpredictability of the world. Indeed, they need to embrace and value it. Expert forecasters, especially economic ones, are incredibly valuable but no one wins if their work is over-simplified or de-humanised. You can never be sure what will happen on a long journey, but you should at least have a map, and economists are often the best cartographers we have available.

Scientists took what the ancient astrologers got right; the patterns and the movement of planets and stars, the maths that predicted their paths, and discarded the idea that they might influence human affairs. Astrologers lay the foundations of what would become a scientific, human understanding of the universe. Perhaps economists' work – analytical, honest, rigorous, insightful, philosophical work – will form the basis of something else. Something more accurate, more radical, more accessible. Freed from the weight of having to predict incidents beyond its capabilities, economics could pave the way for future experts, even scientists to take up their good work and cast aside the unreliable assumptions.

Such a new discipline would almost certainly require even more input from technology, the sector increasingly underpinning every aspect of government and business. Too much faith can be placed in technology, especially where irrational humans are concerned, but it is inevitable that those in charge will see it as the potential solution to prediction.

15
PREDICTING BEHAVIOURS

If economics failed to accurately predict, and sometimes even to explain, major economic events, might it be helpful to gain a greater understanding of the humans that sustain it all? Could those elements that undermine self-interest, and therefore a great deal of prediction – elements such as empathy, peer-pressure, and self-delusion – improve matters? Fundamental to many of economics' failings is the rational actor, as examined by Rational Choice Theory. If psychology had influenced how business made a profit, perhaps the answer to predicting human economic behaviour accurately lay in that direction.

By the mid-20th century medical science and psychology were working together in a new field; neuroscience. A field only made possible with advances in technology as well as other scientific developments, it aimed to understand the brain on a molecular, even atomic level. Its starting point was that memories, thoughts, emotions, impulses – all areas of concern for psychology - must also have some level of physical manifestation. Something we all share, to some degree or another, regardless of the influences of environment and experience studied in psychology. The chemistry and physiognomy of the brain would provide new insights into why people did what they did.

Neuroscience discovered that our brains would act in ways that were largely universal and over which we had little or no control. Chemicals released under certain conditions, often related to primitive survival instincts, changed how we felt, and consequently reacted, on a fundamental level. It revealed why the approval of others, why helping others, and why social acceptance were all felt to be positive experiences regardless of who we were.

After the Second World War many academics studied the causes and effects of what had happened during modern humanity's darkest period. Of particular interest was how an entire nation appeared to undergo a collective psychotic episode. New York psychologist Stanley Milgram was one such academic. He wanted to understand what had made so many people so apparently willing to participate in, or at least ignore, the horrors of the holocaust. Surely not all of the people, the thousands that were needed to carry out such organised horrors, were all somehow evil or sociopathic.

Milgram devised a study using electric shocks; or so the ordinary men and women who volunteered thought. The volunteers were separated into 'teachers' and 'students', but they were unaware that the students were in fact part of the experiment. Under instruction and observation by an academic, the teachers asked the students a series of questions. Whenever they failed to answer a question correctly, the teacher was ordered by the academic to administer an electric shock to the student. In fact, there was no shock; the students feigned their pain, even pleading for mercy. The teacher volunteers reported experiencing anxiety and stress and they objected verbally, but the clear majority increased the voltage when told to and kept shocking. They gave up their sense of morality to a figure of authority, in this case the academic carrying out the experiment. All the volunteers did was obey authority. They were not psychopaths or mentally ill in any way.

With this Milgram demonstrated that people were as influenced by social expectations and their place in the order of things as by any upbringing or hereditary aspects. For another experiment, his students went around the New York Subway asking people for their seats, with no explanation. Many people gave up their seats, without question. But in fact it was the difficulty that the students had in asking in the first place that was more telling. That sense of awkwardness, even stress, was the real power.

Less than a decade later, in 1971, at Stanford University, a group led by Philip Zimbardo undertook a military funded experiment. A set of students were randomly allocated into two groups: 'prisoners' and 'guards'. The latter were given uniforms, batons, and sunglasses (to restrict eye contact) but could not physically harm the prisoners, who were also uniformed and only referred to by numbers. Within just a couple of days prisoners had started to resent and revolt against the guards; whilst guards displayed authoritarianism, even sadism, and tried to sow division amongst the prisoners. Although much challenged and questioned since, the experiment has become one of the most famous social psychology experiments ever designed, and has inspired a great deal of work on social pressures, conformity and behaviour.

Social influence plays a significant role in money –

how it is spent, earned and exchanged. Money demonstrably inspires irrational behaviour. It only has value because most of humanity agrees, without any debate, that it has value. That is irrational enough, but its effects can be even more illogical.

Once again, Adam Smith was a visionary. Alongside ideas around the value of work he also considered the effects of money on concepts of fairness and morality. Jeremy Bentham looked at how happiness could be valued as well as maximised. They both operated in a time before psychology was formally recognised. With widespread acceptance and greater understanding of psychology, some academics inevitably examined the effects of psychology on money and vice versa.

In the early 20th century Irving Fisher, he of the statements of pre-Wall Street Crash confidence, was one of a number of economists applying the new ideas of psychology to economic studies. The development of Game Theory was another example of psychology making its presence felt in the field of economics, demonstrating that humans may not always be the rational actors they had been assumed but that when it came to trust more personal and social influences came into play.

By the 1960s some psychologists were specialising in the field of money, value, risk and decision-making. Most notable were the contributions of University of Chicago cognitive psychologists Amos Tversky and Daniel Kahneman.

Cognitive psychology focuses on our understanding and use of language, logic and problem-solving. Tversky and Kahneman looked at how their work on decision-making, risk and rationality compared with the accepted economic ideas in these areas. This combination of psychology and economics, specifically the psychology of economic decision-making, became known as behavioural economics.

Behavioural economics looks at the psychology and neuroscience of risk, trust and decision-making, specifically regarding financial matters. By looking at such decision-making, it also infringes on other areas of life like work, personal relationships, and physical wellbeing. Tversky and Kahneman, both alumni of Stanford University, developed experiments and models to more accurately describe the truth behind irrational human behaviours in all of these areas.

The pair looked at why people appeared to struggle to make analytical, dispassionate judgements about risk and probability. Why people so often ignore readily available information to make the sort of rational choices that economists, and others, long believed they would. Given that many financial decisions are, in the short or long-term, based on risk it could have profound effects on understanding a range of business and wider economic issues. They looked at the perspective of the average consumer, and also of those in key positions to influence bigger issues.

Central to Tversky and Kahneman's work was the idea of heuristics, in which people develop largely subconscious shortcuts. The practical upshot of which means human tend to focus on just one, simpler aspect of a problem or choice and resolve that element as a means of resolving the whole thing. That could be anything from choosing between two political parties based on just one or two easily understood positions, or picking which car to buy based on the fact that it is the same manufacturer as your old one. Related to this are a number of biases that, just like the Confirmation Bias, prevent or override more rational judgements, such as the Optimism Bias and the Anchoring Effect.

The Anchoring Effect sees us unduly influenced by irrelevant numbers, especially when it comes to estimating the value of something. Rather than assessing the information dispassionately, we are easily swayed by numbers that we believe to have some bearing. If asked to guess the price of a phone, if you are shown nothing but expensive phones, you may say £300. But if the only comparisons you have are cheaper, you will probably say £150, for exactly the same phone. Very simply, humans are swayed by irrelevant information if that is all that is available. The same can be observed in any numerical estimate and it can in turn affect people's decisions based on what they believe the majority appear to be choosing.

The Optimism Bias, Tversky and Kahneman suggested, may be the most influential of all of the cognitive psychological biases that the pair examined. It describes our tendency to assume, regardless of the statistics, that we will not be subject to bad things; that we are simply at less risk than others. Humans do not deal well with complexity so they rely on those heuristic shortcuts to reach the conclusion that, contrary to what others experience or what the information suggests, we will be OK. This extends well beyond financial risk and into areas like people leading unhealthy lifestyles not thinking they will be susceptible to a heart attack - although this also demonstrates our difficulty in thinking long-term and our desire for immediate gratification. This phenomenon was widely observed in the 2007-08 financial crisis where individuals involved in high-risk trading believed they were less exposed to potential disaster than others. This bias demonstrates the fundamental lack of rationality in human decision-making and our unfounded belief that we have some say in what happens to us.

Whilst this had the potential to have a significant effect

when analysing groups, it could be much more significant when it affects those making important decisions; decisions that could matter to thousands, even millions. Whilst everyone, to some degree, displayed these biases and irrational behaviours, their effect was often moderated. The scale of the individual decisions was often small, or they were diluted by the influence of others.

This was particularly true of those displaying extreme versions of these behaviours. Unfortunately, however, is was clear that, much as humans might like to think their leaders were well-selected and better than them, more rational and clear-minded, more utilitarian than them, it was evident they were not. The Confirmation Bias, the Sunk Cost Fallacy, the Overconfidence Bias (placing undue weight on your own knowledge) were not only evident amongst leaders, they had been reinforced by their success in their chosen field.

Once again, the 2007-08 financial crisis provided an example of how this could cause problems. Perhaps because leaders were now held to more public scrutiny, it became clear that few financial and political leaders had sought to mitigate the effects of these biases. What made it worse was that many in these areas were required to make decisions quickly and under stressful conditions, where the irrational seemed to have much greater influence.

At the same time as Tversky and Kahneman were doing their psychological work, the physiology of the brain was being examined. Different parts of the brain were observed to work in essentially different ways, reacting faster, and stimulating the production of hormones the effects of which were hard for an individual to resist, or even be aware of. It was an examination into what had always been considered to be human nature.

At Stanford, psychologist Walter Mischel defined Hot and Cool Cognition. His most famous experiment took a group of four-year-olds and placed them individually in a room with a marshmallow on the table. They were told that they could eat the marshmallow, but if they resisted for 15 minutes, they would get two marshmallows. Naturally some resisted, whilst some went for the instant gratification. Mischel followed the children's progress through education and adult life. He noted that those able to resist demonstrably performed better in school tests as well as being generally healthier and even happier overall. Mischel termed the impulsive, quick to consume behaviour Hot Cognition.

The idea of Hot and Cool decision-making was popularised decades later in Kahneman's bestselling book *Thinking, Fast and Slow*, in which he summarised much of his and Tversky's work. They proposed that humans have developed two decision-making systems: System One - the impulsive, unconscious, irrational one; and System Two - the deliberate, conscious, rational one. The two worked quite differently to produce different conclusions. By understanding this, we could start to know when one system is over-riding the other and therefore avoid poor decision-making, or at least poorly reasoned decision-making. It was this examination of biases and decisions that made *Thinking, Fast and Slow* so popular in the years after the financial crisis.

That many decisions, even quite important decisions, were frequently more impulsive than rational explained a great deal, and challenged classical economic assumptions. Whilst people may well be, at base, self-interested, they are not always equipped or able to make rational, and as such predictable, decisions. But what if the process was wrong, but the outcome was right? Somewhat counter to the work of many in the field of behavioural economics, academics such as Gert Gigerenzer suggested that, in attempting to mitigate the risk of bad things happening, we might actually make the whole situation worse. This might be particularly the case when we do not fully understand the risk – its nature or its size. Instead, we only take information from limited sources; potentially biased or out-of-context sources, but nonetheless we assume it is usefully informing our decision. After all, we do not know what we do not know, but we make what we believe to be an informed decision anyway.

Gigerenzer concludes that people are not stupid, and are not as misled by instinct as might be assumed. Particularly in the modern age, with so many variables and sources of information, it is almost impossible for someone to make a truly informed decision. So we must do the best we can and, on balance, it works out well. The occasional error or misjudgement just leads to new learning, new adaptations and better future decisions. The assumption that many people make poor decisions is potentially dangerous. It could be used to justify increasing authoritarianism, or a greater reliance on being given fewer choices; of being pushed towards choices 'for our own good'.

We all build up our wealth of experience and knowledge and the balance of probabilities is that we will make the right decision when called upon. This is particularly the case in collective decision-making, provided a varied enough cross-section of experiences and knowledge is consulted. Generally, the biases, impulses and understanding even out. In many cases, making a decision is all that is required. Voltaire popularised the phrase "the perfect is the enemy of the good". That is to say, in searching for a perfect decision, a good one is overlooked, or possibly, none is ever reached at all. A bad decision can often be useful and instructive, leading to new destinations, engendering new skills. A good decision is a good decision. No decision is almost always the worst option; although, in an age of access to more information than anyone could possibly hope to consume, no decision is an increasingly tempting option.

Sport is often cited as an example of instinct outperforming analysis. Whether it is a footballer measuring a pass across the field or a baseball player positioning themselves, they tend not to calculate wind speed, parabolic curves, friction coefficients and gravity (as a computer, or perhaps an economist would) in order to estimate where they are best placed in order to reach their teammate or make the catch. Thousands of failures and successes, of watching others and simply knowing their job means, whilst they will not be right every time, for the top performers it becomes instinctive, and they will be right more often than not. Reactions rather than thought process gets them to the top.

Does this mean the decision-making process is more irrational but the result more rational and consequently predictable than expected? Without being able to fully understand each individual, the outcome, regardless of the process, will still often be uncertain. We all estimate probability and we all do so based on our experience. If we have never been in a car accident, we generally assume every time we get into a car we will get to our destination unscathed. But if we have, we may be slightly more cautious. In fact, we never really consider the actual statistical probability, and we certainly do not plan our route to ensure the safest possible journey, and in many cases we never could.

Rational Choice Theory, that cornerstone of economic behaviour, also focused less on the process and more on the outcome of the decision. That regardless of how an individual arrives at a decision, it will still reflect self-interest or will prioritise utility at some level. Either way, the model of the rational economic actor remains unreliable.

We hold unconscious biases and are impulsive. We form and inherit concepts of fairness. We are unable to think long-term or control the effects of anxiety or of euphoria on our decision-making. We place status or an emotional attachment on inanimate objects. No one can predict unintended consequences. To be rational we would have to re-evaluate our emotional connection to money, statistics and other humans. How we measure our own worth and our decisions. How we use money unwisely to justify a wide range of actions or to avoid admitting we were wrong.

In the mid-1990s initial work was undertaken on understanding how psychology related to motivation. It considered that, if you ordered someone to do something, they were quite likely to ignore the instruction or react against it. Alternatively, explain it to them, and risk them reaching an irrational decision of their own anyway. It might be that a psychological, subconscious way of changing behaviours would be more effective.

Behavioural economics reveals, amongst other things, the mistake of conflating self-interest (which can include everything from laziness to peer-pressure) with economic self-interest. In fact, the two are often contradictory, especially when considering the long-term. Adam Smith concluded that economic self-interest is the force that creates the most efficient market. The free population will unknowingly divide up the money and the work and allocate the two in the most effective way. In this, the market will always be rational and right, even if individuals are not. Unless government, which is often just as irrational as the people that elect it, gets it wrong.

One of the chief problems of democratic government is that its self-interest - that is, the desire to stay in power – is not always congruent with national self-interest. Especially when it comes to the long-term. Were the two interests the same, all governments would, by and large, divert almost all of its resources into education. In Adam Smith's and Milton Freidman's terms, it would focus on the Human Capital upon which all other capital depends.

Along with the economic benefits, almost every available study shows a good education system sees a healthier, happier, more productive population with less crime and inequality. Yet politics does not put all of its efforts into education. Worse, it tinkers with it. It imposes quick, superficial, and ultimately worthless changes because voters demand quick, visible actions. Robust education policy takes years to form, needs constant review, take years to implement and even longer to produce tangible results. By which time the politicians responsible for it have lost power and quite possibly left politics, often for a career in the private sector. As a result, education has to compete with many other areas of public spending, it is under-resourced, and a market in education develops whereby the wealthy (including many politicians) can buy an advantage.

Behavioural economics would not just address the shortcomings of classical economics by explaining biases and poor decision-making. It would evolve into more than just a theory; it would become a practical process. Knowing that people are subject to biases, and that they struggle with long-term decision-making, could they be corrected? Could an outside authority, a benign one with individual's or communities' best interests at heart, use psychological understanding to push them towards a particular choice or behaviour? Could they be made to be more self-interested? More rational; more predictable?

At Chicago University, Tversky and Kahneman worked briefly with another academic, Richard Thaler. Along with law professor Cass Sunstein, Thaler took behavioural economics a step further. In their book, *Nudge: Improving Decisions About Health, Wealth, and Happiness*, Sunstein and Thaler looked at the barriers and incentives involved in making good and poor choices and decisions. Furthermore, they looked at real world examples of how it could improve the world.

They considered 'choice architecture' in which different ways of presenting choices can have significant influence over the outcome. In the same way Tversky and Kahneman had studied how a simple change of language or context could see dramatically different interpretations. For example, they presented people with 600 theoretical patients all suffering from a fatal disease. Asking participants to choose between Treatments A or B, some were told that A will save 200 people, and some were told that A would see 400 people die. 72% of respondents chose A when framed in the first, positive context as opposed to 22% when framed negatively.

Thaler and Sunstein's book looked at why people act in the way they do, and what the implications were for those concerned with human wellbeing. If you want to motivate people to make the choice that, dispassionately, is best for them, then you need to frame it appropriately. They need to be subconsciously 'nudged' towards making the right choice. If you must give them an option, frame the 'right' one in a positive way. Between offering a discount or a fine (even if it amounts to the same thing), the former will be more effective. Better still, make the option you want them to choose, the one that is best for them or their community, the default. Make them work or go out of their way in order to choose the alternative option that you would rather they did not.

Nudging takes advantage of System 1 or Hot thinking. It actively uses the impulsive, irrational side of thinking, the part that uses the shortcut when there is too much complexity, to create questions or environments where people do not think about the choices they have; they just act. Only nudging means you can be fairly sure they will act in the 'right' way.

The idea is that people, for example, want to be healthy, but irrational decisions get in the way. The desire for instant gratification, their belief that they will not be ill, that they can compensate tomorrow for poor choices today, and all the other biases that behavioural psychology uncovered. Smith's, and his successors', belief in self-interest was being undone by the inconvenient truth of human nature.

Another aspect of human nature is that we tend towards inaction; we seek a type of physical and mental equilibrium. The greater the 'friction' we have to overcome, the less likely we are to undertake an action. Again, this friction can be used to nudge people towards a particular choice.

A commonly used example is that of gym membership, which once signed up for, is rarely cancelled. This is partly because people fool themselves into thinking they will go to the gym; just not today, so it is best to keep the option open. It is also because it requires effort, and gym owners know that the harder they make it, or at least the harder you think it will be, the less likely you are to cancel. Even though you are throwing money away every month. Even though you know that money could buy healthier food, a bicycle, or a rowing machine. Even though you know you are being irrational.

Experiments have shown that we will inconvenience ourselves to quite some degree to save 50% on something that costs £10. But we would not do anything to save 5% on something that costs £100. If we lose £100 on a bet or a bad investment one week, but the next week we make £150, we feel much more aggrieved over the loss than we do elated over the gain, despite being £50 up on the fortnight. The gain usually has to be at least double the loss in order to compensate.

There are so many examples of this type of irrational behaviour, especially when it comes to money. We will spend more on something others can see than on items hidden from view, even if the latter would benefit us more. We are more likely to offer comments if we are asked to help improve something rather than being asked for our feedback. We are subject to social pressures if we believe many others are spending in a particular way. We place undue emphasis on advertising and brands, even if they are for substandard products. Time and again we are seen to be more likely to spend more on the same item if we are paying by card rather than cash. And even more again when we are paying by subscription or some other 'frictionless' type of payment such as that used by Uber. Yet it is the same money. Our bank account, our employer does not distinguish between one type of spending and another. It is all money, but in our minds, it is very different. It is all quite irrational and it is all quite instructive for those that would exploit our lack of logic.

Nudge theory found a ready home amongst Western

governments as well as in business. It helped solve some of the frustrations of having to deal with large groups of irrational humans. It shaped policy on pension schemes and organ donation, both of which made use of the friction principle. A number of governments established so-called 'nudge units' with over 80 countries actively employing nudge-style policies.

Nudges have the benefit of usually being simple and relatively cheap to implement. In the UK, nudges (or behavioural insights) have seen the limiting of the number of takeaways near schools, as well as the widely publicised hostile or challenging environment policy for immigrants. This latter application saw government agencies encourage (or nudge) immigrants to leave the country, not directly, but by making their day to day lives harder by degrees. It was kept largely secret within government and the civil service until investigative journalists revealed that people, even those with long-standing permission to stay in the country, were denied hospital treatment, jobs and housing.

Classical economics relied on rational, informed, self-interested choices for its predictions both on the large and small scale of the consumer and the national economy. But psychology and behavioural science suggested that it is rarely that simple. People are impulsive, especially when faced with complexity or stress. They do not have access to all the facts, and even when they do, they only pick out the bits they like or agree with. Humans working solely to their own advantage, if it was ever true, it certainly was not anymore.

Psychology and neuroscience provided a much sounder foundation from which predict the behaviour of both groups and individuals, and also to indicate when prediction might be impossible. It accounted for charitable giving and ethical consumerism, but it also accounted for impulsive, pressurised and uninformed decision-making.

If economics had failed to predict financial crises, would behavioural economics provide assistance? Could this analysis of an individual's irrational behaviour be useful when it came to both the decisions of the average consumer, and that of investment bankers, commodities traders, company boards, politicians, hospital administrators and others? Or perhaps nudge theory provided a way of making them more reliably predictable and avoiding the problems of the past? Or is this sort of prediction more like physics, whereby the methods used at the very small, sub-atomic or individual level are entirely different to those used at the big, planetary or macroeconomic level making the holy grail a unified theory of prediction?

As well as governments seeking efficient ways to encourage populations to save, be healthier, use less energy and commit less crime, behavioural economics and nudging found another welcome home. A place always ready to embrace new ideas. Somewhere that had for over a century been quick to find ways to commercialise the new, from railroads to broadcast media, creative arts to behavioural psychology. Home to pioneers of all sorts: California.

16

AT THE FOREFRONT OF TECHNOLOGY

The Industrial Revolution saw the start of the geographical gathering together of related industries. This was particularly evident in port cities where businesses that needed to import and export moved closer to sources of labour, finance (in the form of investors and banks), ideas (educational institutions), influence (government and administrators) and other suppliers (such as the makers of new equipment). Economists called this clustering of interdependent yet independent organisations agglomeration (derived from the Latin meaning to gather together in a ball or mass) and it was visible across the industrialised world.

For centuries a ship would be loaded and unloaded almost entirely by hand. Men would load pallets, barrels, boxes and crates into the hold of a ship. At the destination port, men would unload. Each time, even with the use of cranes, only small amounts could be moved. It was hard work, employing many men working long hours, and rarely would a week pass on a busy port without at least one fatality. Advances in pulleys, nets, pallets and so on had only limited improvement.

In North Carolina in 1935, a young man called Malcolm McLean left school with good results but few opportunities. Unable to afford to further his education, his family scraped together enough money to buy a truck. With his siblings Clara and Jim they started the McLean Trucking Company. The state had long been at the heart of the US tobacco industry and Malcolm was driving empty tobacco barrels around. The business grew along conventional lines of hard work and small profits.

During the Second World War, the US would ship fully-laden trucks to Europe. McLean put this idea into commercial practice, shifting trucks rather than just their cargo to New York by sea. This was an advance, but it meant filling the ship's cargo hold with trucks – chassis, cabs and all – and that was not efficient. He looked into designing a metal box that could fit onto a truck, be lifted off, and then to be placed directly onto a ship.

There followed a typically American entrepreneurial story of never giving up and finding opportunity in what others saw as problems. McLean exploited a legal loophole in the rules banning ownership of a trucking and a shipping company - designed to prevent too much power resting in too few hands. When dockers, seeing what McLean's development might mean for the future of their work, went on strike. In response, McLean spent the downtime refitting two wartime tankers to carry his new boxes. In 1956 he launched the converted ships. At the time, loading a ship cost around $5.86 per ton. Using what became known as containerisation reduced that cost to ¢16 per ton.

McLean changed the face of shipping and trade. There consequently came a change in employment, wealth, and manufacturing, as well as the social and economic make up of the port cities. McLean's company Sea-Land opened up routes around the world. The company took on the US government's contract to ship to and from Vietnam during the war. During which, McLean realised that he needed to be shipping something back in those empty containers being returned to the US. The company would stop-off at the then growing consumer powerhouse of Japan. It brought cheaper products, especially electronics, back to America.

Sea-Land sought standardisation within the shipping industry in order to lower costs even further, and McLean released the patents for his container designs for royalty-free use by anyone. Competitors moved to this type of shipping and soon all ships were fitted to take a standard 40 or 20 foot container and ports had the cranes and spaces to move and house them.

Containerisation became so efficient it is estimated the total cost of shipping a ton of goods from Europe to the US is now around $50. Businesses basically assume that there is no cost to such shipping anymore. It has gone from consuming around a quarter of the cost of a product or commodity, to being negligible.

As shipping became cheaper and less labour intensive, so the port cities changed, possibly quicker than any urban area had changed since the Industrial Revolution. Once a ship would be in dock for days, sometimes weeks. Cargo was unloaded, accounted for, separated, allocated a warehouse and then replaced by more goods. Scores even hundreds of crew would be in port, spending and working during that time. Today a ship can be in port for a few hours, a day at most, before making its onward journey. Crew sometimes do not even disembark from their ship.

Trade built ports into vital economic centres filled with warehouses, financiers and merchants. The Industrial Revolution brought the rapid expansion of port cities. Transporting goods over land was slow and expensive, so it made sense for manufacturers to be as close as possible to a port – near to raw materials coming in, near to the point of export for their finished goods. Where once factories were built wherever land and labour were cheap and available, now they were built close to docks. It made cities like London, Hamburg and New York centres of every conceivable type of business. In addition to the financing, purchasing and sale of almost anything, with so many ships' crews around for long periods, economies catering to them and their lives also emerged.

Containerisation changed all of this completely. The desire for cheap land drove factories away from expensive, crowded cities now that transportation costs were much lower. There was no longer an economic advantage to being close to the point of import and export - even to the point of multinational supply chains becoming commonplace. Ships grew and hundreds of tons of cargo could be loaded or unloaded in a day. Docks inevitably moved out of cities as their need for more space made expansion impossible or too expensive. The old docks died. To be replaced by what? First wasteland, but then, as the nearby city drove up land prices, redevelopment. Skyscrapers replaced cranes; luxury flats replaced warehouses.

As Adam Smith predicted specialisation was the way to

an efficient, if not always humane, economy. It enabled others to become expert at specific jobs, which meant cheaper, more efficient production. With large-scale shipping costs now negligible, the growth of international supply chains meant ever more specialisation as one factory could now supply a single, niche item to scores of customers around the world, not just the ones geographically closest. Cars, an example of a product requiring hundreds of individual parts, were no longer built in a local factory but simply assembled; metal, glass and plastic jigsaws with individual pieces shipped from all over the world. A maker of windscreen wipers in China, a maker of mirrors in Brazil, a maker of locks in Poland, all able to ship their particular specialisations to car plants in America, Britain, Spain and Korea. All this because one man started shipping goods in a big metal box.

With the trade in physical goods now mastered, a new trade became the greater source of profit. With cheap land and labour more abundant in developing economies and poorer countries; the manufacture of physical products shifted more in their direction. Meanwhile ideas, advice, culture, education became the basis of Western economies. Dull, dirty and dangerous work was left to others. The rich countries focused on entertainment, law, advertising, architecture, science, design – intellect and services. Work that involved very little physical product. There was no need for the physical labour of thousands of humans. No need to keep trying to find ways to replace that labour with better machines. No longer any inefficient packaging and transporting.

Containerisation played a big role in diluting industrial agglomeration. As transport was cheaper, shifting factories and warehouses to where land and labour was plentiful (and therefore cheaper) geography became less important. However, the shipping of heavy products over distances was not the only reason for companies to group together.

Throughout the 20th century, those requiring particular skills, finance, or ideas gathered geographically close. Agglomeration, however, comes with problems. As new groups of connected industries gather together land becomes more expensive. Residential property can be scarce, creating congestion and pollution as people travel to work from cheaper areas. Commercially, this geographical accumulation of businesses can see cartels of established companies push newer, entrepreneurial ventures out by limiting their access to suppliers and workers, or it can see those bigger players demand fealty from newcomers. Despite this, it remains a fertile environment for economic growth; as relevant to the new business of non-physical trade as it had been to the physical.

HOW THE WEST WON

Europeans first explored California in the mid-16th century but for many years it remained remote, difficult to access by land - across the whole continent of North America and through mountainous and desert terrain; or by sea - across the vast Pacific. It was sparsely populated with small trading posts and settlements inhabited by only the most hardy of people. It was part of Mexico for the years immediately after that country won its independence from Spain. Its early economy relied almost exclusively on cattle, seeing ranchers from across Mexico, America and Europe grazing cattle in a region that largely resisted attempts at being governed by any one power.

After a period of conflict, California became the 31st state of the Union in 1850, but still only sustained a population of a few thousand.

James Marshall, a New Jersey-born farmer and carpenter, travelled across America, setting up farmsteads and moving south on the advice of doctors. He settled in California in 1845 and, having joined the military fight against Mexican authority in the region, returned home to his farm to find it all but destroyed. Undeterred, he and a fellow settler and war veteran set up a sawmill on the American River. In 1848 Marshall noticed something in a water channel by the mill. He collected a few pieces of a shiny material, examined it, and with a basic knowledge of minerals concluded it could be nothing other than gold.

News of the discovery of gold, thanks in part to the new telegraph system, travelled quickly around the world. The California Gold Rush saw 300,000 people (referred to as forty-niners – arriving in the year 1849) from elsewhere in the US and from around the world flock to this new land of opportunity. As well as those seeking gold, there came those seeking to supply them with tools, shelter, provisions, diversions and security. Plans to build railroad and telegraph systems to connect the state were prioritised. Agriculture expanded to feed the exploding population. Whilst much of the gold discovered (billions of dollars in today's terms) went to already wealthy individuals able to pay workers and buy the best equipment (and politicians), the idea that anyone, if brave and diligent enough, could become rich inspired thousands.

California, especially in the new gold-centric settlements, was a lawless place. Still remote, it was often beyond the legislative reach of Washington and was, between 1848 and

1850, neither American nor Mexican, but under military rule. This meant that land, and the gold or anything else on it, was seen as free for the taking. It was not just a case of staking your claim; it was also about protecting it. By any means necessary. More scurrilous practices aside, the generally the agreed rule was, if you worked your claim, it was yours.

The myth of California developed a life of its own; a myth that would see its governance play a key role in the Civil War. A myth that would attract thousands more people during the Great Depression. Its rapid economic expansion affected both the American national economy, and others around the world. Its reputation as a place for a new start, of opportunity, of where anyone, no matter who they were or how humble their background, could become great would define its narrative for the next 150 years. Gold made California rich, and wealth attracts more wealth; companies and individuals seeking to tap into affluent new markets or to make a commercial killing before their rivals. Ironically, that which made it rich – gold - was also a sign of its weakness – the state's perilous position on the major tectonic fault which created those seams of precious metal thousands of millennia earlier. But even that fed the Californian narrative that great wealth came at great risk.

After gold came oil - and where gold had established San Francisco as the major Californian city, oil saw Los Angeles compete with it for influence. The space and reliable weather on offer, as well as the new money, saw innovative industries move to the state. Industries such as aviation, military technology, and, of course, the entertainment industries. The busy port at San Francisco also saw early telegraphy and radio innovations born there. The youth and relative isolation of the state also inspired a strong sense of regionalism; an opposition to the perceived sense of superiority from the east that would try to take or exploit California's wealth and talent.

THE GOLDEN STATE

Just as Britain led the Industrial Revolution through a combination of Enlightenment ingenuity, good fortune, and internationalism, so America's lead in technology grew from similar seeds. A nation built on immigration and openness to the world; that welcomed many seeking refuge from postwar Europe and conflicts around Asia and the Middle East. A nation built on an unshakable belief in commerce, industry, individualism, and pioneering risk.

Europe, for all its advantages - for Germany's engineering, France's creativity, Britain's industry - spent too much time arguing and fighting. It looked more at its differences than its similarities and shared goals. America, a diverse continent of many differences, unified under one language, flag and constitution. It became a global magnet for people and ideas, nurturing collaboration and growth. It was an inspiring idea; a land of opportunity.

The belief in the US and its people as being a dynamic, pioneering nation acquired a mythical, and passionately held, status. A young nation has little history, but tales from Lewis and Clark to the Wild West ensured that what America lacked in historical narrative it made up for in stirring, dramatic stories of bravery and exploration.

America had pioneering embedded into its being. A new nation, sometimes impetuous, but always willing to take risks. A nation, a project, a clean start having learned from the mistakes of older nations. A vibrant, agile nation; a bastion of democratic, free-thinking (or at least it felt it was). It did not always work, of course; failure is inevitable and even necessary. But its determination to build and to populate the vast, varied American expanse meant allowing people the freedom to make their own rules.

Throughout the early 20th century, improved transport, first by rail and then road and later air saw California grow from under one million people to the most populated state in the Union by 1965. Part of accommodating this rapid growth, especially in a state that was hard to access and largely bordered by desert, was major infrastructure projects. This climate of growth, funding and innovation meant a demand for the best engineers, scientists and technicians. That demand in turn saw the likes of Stanford University and California Institute of Technology (Caltech) became world-leading institutions of research and education.

The City of Palo Alto was built by Leland Stanford Sr. The son of a farmer, he studied law, entered politics, and moved to California during the Gold Rush. He and his brothers set up a successful general store and warehouse business and he went on to become Governor and Senator of the state. He also founded Stanford University, around which Palo Alto grew. The University struggled financially for many years after the death of its founder, and either side of the war its Provost and former Dean of Engineering Frederick Terman encouraged students to pursue entrepreneurial projects for the financial sake of the institution, the local economy, and for their own education.

In 1937 two former students of Terman's, Bill Hewlett and Dave Packard scraped together just over $500 and, in a garage/shed in which Bill was sleeping, the two went into business. Hewlett-Packard's first product was an electrical oscillator which produced an audio signal and that found an early application in the testing of cinema sound systems for Walt Disney Studios. The partners were recruited by the military during the war and they worked on munitions and radar technologies, giving them both funding and access to the latest research.

The Second World War saw California become a major manufacturer of military equipment, particularly naval vessels with its location on the Pacific and good natural harbours. The proximity of flat, open plains had seen one of the first military airbases built in 1933 and from there a growing, cutting-edge aerospace industry. In 1939 one airfield, Moffett, opened what would become a leading research base for jet engines and later one of Nasa's first homes, the Ames Research Centre and Jet Propulsion Lab.

With funding, training and facilities, both public and private, California became home to some of the key advances in science and technology. A significant amount of work on military communications technology during the war led to experiments into silicon and its behaviour as a conductor of electricity. It was a complicated physical process, but by the late 1940s silicon (primarily, but also other semiconducting materials) had been adapted to create transistors, a form of electronic switch. Arrays of these switches could control any number of devices and, making use of quantum physics, even store information.

By the late 1950s semiconductor research represented a brave and exciting new frontier of science and technology. Many involved in it believed it had the potential to change everything. These were pioneering times in a pioneering place and semiconductor research flourished in a greenhouse of military, scientific and commercial progress. Caltech graduate William Shockley, co-inventor of the transistor, started the first large-scale commercial semiconductor business. In 1957 eight of his employees (referred to by him as the 'Traitorous Eight') started the rival Fairchild company.

Amongst that Traitorous Eight was another Caltech graduate, Gordon Moore. In 1968 he co-founded Intel, but three years earlier he made a prediction that would become known as Moore's Law. He predicted that the number of transistors that could fit onto a piece of silicon would double approximately every two years (although others have later quoted his timeframe as anything from 12 months to almost three years) for the next decade. This rule of thumb would come to underpin the exponential rate of growth of microelectronics (the doubling of speed, capacity and processing power and the halving of costs) that would feed the digital revolution 40 years later.

Fairchild became the first and for a while only company able to make reliable, working transistors. They received significant government funding, in part to work on the nascent US space programme. The programme demanded rapid technological advances in order to overtake the Soviet initiative that was threatening to run away unchecked.

To further fuel the fire of innovation Nasa moved into the area in the late 1950s with its acquisition of the Ames facility. A government body, civilian and scientific in nature with a distinctly military influence, it combined brilliant minds from all over the world and advanced science and technology in incredible ways. Although sometimes seen as a propaganda tool in the Cold War, their work had great scientific value as well as military and commercial applications. In addition to exploring and advancing our understanding of the universe, they laid the foundations for future private money-spinners from foam mattresses to cordless tools to smartphones. They did not just discover; they helped build an environment that enabled and encouraged further discovery.

It was a dynamic period and the scientists and engineers of the Palo Alto area were maverick, brilliant and often misunderstood. Traditional businesses often saw them as, at best unconventional, sometimes downright wrong in how they operated. More established companies were often wary of getting too involved, fearful of what they did not really understand and reluctant to change what they saw as a successful way of working. Over time, however, they became more eager to make use of the products the companies created, to the point they came to depend utterly on them.

Whilst Shockley were the first, Fairchild were the most progressive and influential of the semiconductor manufacturers. Their work pushed semiconductor electronics on significantly (Hewlett-Packard remained focused on solid state electronics for some years and only joined the silicon gold rush in the early 1960s). Fairchild's business model of encouraging new enterprises, amongst them AMD and Intel, also established a blueprint for what in 1971 was dubbed Silicon Valley.

The semiconductor transistor gave a new impetus to the world of computing. Previously computing was limited in its practicality, and machines were enormous, slow and unreliable. Semiconductors solved all of these problems. In 1951 the first commercial computer was available and by 1953 IBM (who had by now joined the key players in the Valley) were mass-producing machines. The advent of the integrated circuit, that is a piece of semiconductor containing many components without the need for wires to connect them, made for still smaller and more reliable machines. By the early 1960s the computer as many of us might recognise it was widely available - although not everyone knew quite what to do with it.

The use of computers in the military, and their potential, saw the Pentagon's Advance Research Projects Agency (ARPA) draw up plans for computers to be connected across the country. APRA's director of Information Processing Techniques (and later its Director of Behavioural Sciences and Command & Control Research), computer scientist Joseph Licklider had a vision for departments, institutions and organisations whereby they could all openly share information via computer. His idea inspired the creation of ARPANET, a computer network which connected sites across the US. It started by connecting four sites, three of which were in California including UCLA and Stanford. The RAND Corporation saw potential in this network as a tool in the struggle to survive in the aftermath of a nuclear attack. The military saw the dual opportunity to improve tactics and decision-making whilst also creating a nuclear command and control structure. Academics saw an unprecedented chance to collaborate with those few that had access to such large, expensive machines. Although the first two of these never really came to pass, the military did run ARPANET for its first two decades of existence until 1990. No one, however, foresaw just what its progeny, the internet, would become.

As well as connecting computers, the 1960s also saw the arrival of the supercomputer; large-scale powerful banks of computer processors able to handle highly complex operations and vast amounts of data. Like most developments in the area it came about through a combination of academic enquiry and commercial imperative. The early supercomputers saw application in oil extraction and in the US nuclear weapons programme. Their ability to process vast amounts of data meant their uses were limited largely to areas of scientific research where many measurements were taken, many variables were possible and multiple outcomes needed to be accurately assessed. They were used to model and predict the behaviour of weather and of sub-atomic particles.

In the late-1960s the Valley, now a hub of finance, research and freethinking, was joined by an influx of immigrants fleeing conflict in South East Asia bringing skilled technical workers, trained scientists and engineers, and new ideas and perspectives. By the 1970s the area around Palo Alto and the Santa Clara Valley was home to dozens of labs and businesses. As well as Intel and HP, by the close of the decade Apple and Oracle had joined their number, and Xerox built a groundbreaking facility that would develop an early graphics-based user interface amongst other influential innovations. These were companies that, beyond providing complex, high-end equipment for specialist use in research, government and business, were now providing computers for the home and small business. They were actively spreading the Computer Revolution around the world.

Although Fairchild were seen as the first technology company to court outside investment, the 1970s also saw the development of a new form of finance for these high-tech, high-risk, complicated businesses. Their rapid growth and need for large levels of early investment in equipment and people, added to the fact many were started by less commercially-aware founders, meant traditional investors were wary and a burgeoning venture capital industry developed around the Valley, specifically in Sand Hill Road. Venture Capital (or VC) specialised in assessing which small, early-stage businesses would provide rapid growth, and as such big returns. VC would go on to play an increasing role in the establishment of many Valley businesses.

The 1970s saw Silicon Valley not just gain its name, but also an unstoppable momentum. The decade saw the first artificial intelligence research at Stanford and the first biotech company move to the state. It heralded the invention of fibre optics, computer-designed graphics for film, the floppy disk, video games, assorted programming languages, and the computer mouse, amongst many other more esoteric, but nonetheless vital developments.

Each advance, and the research that underpinned it, was driven by a sense of scientific and technical discovery. As well as a desire to push the boundaries of what was possible, there was, of course, personal and commercial rivalries. The Valley was a microcosm of the American economy and of America's place in the world, particularly during the Cold War. Risk-taking individuals were taking on convention, embracing the new, creating their own stories, making money and changing the world.

IDEAS AND IDEALS

All of this technical and commercial innovation and dynamism did not occur inside a bubble of industry and academia. Culturally California in the 1960s and 70s was awash with new ideas and lifestyles. The film, television and music industries that had made Los Angeles their home brought creativity, youth, idealism, rebellion and counter-culture – along with a large dose of commerce. Most prominent of the fads, subcultures and cliques that developed in the decade or so after the so-called birth of rock 'n' roll and the teenager, was the hippie. Although relatively brief in duration the hippie movement was surprisingly influential, nowhere more than in California.

Although by nature hard to define, the hippie philosophy brought together many old and new ideas about the world and the idea of self. In what many saw as self-indulgent, its amorphous nature saw it taking and adapting elements of paganism, environmentalism, individualism (including some concepts from Nietzsche about society suppressing people's potential), Asian mysticism, socialism, astrology, and pacifism, and combined them with drugs and music and the fear of global nuclear Armageddon. Its focus on challenging conventional notions of society, perception and thought, as well as its lack of a single narrative or central text, meant many viewed it with suspicion, and often, outright hostility.

At its core, hippie culture meant a sense of freedom – freedom to think, to act and to 'be'. It popularised notions of 'finding' and understanding oneself; of putting yourself first, understanding your self-worth and your potential, and freeing yourself from the stress and suffering largely derived from peer pressure and societal expectations.

Similar to Buddhism, the religion it most closely, albeit it distantly, mirrored, hippie culture readily incorporated many ideas but rarely prescribed any of them. Its focus on the self meant the freedom to adopt philosophies almost at will. Whilst renouncing materialism, it was flexible enough for the wealthy and ambitious to embrace it. Critics belittled it as a woolly, pick and choose ideology, happy to include the best and the worst of human nature. It was accused of being a deliberately ill-defined lifestyle choice filled with hypocrisy that did little but offer a temporary home to those wishing to rebel against comfortable, conventional middle-class upbringings. Those that can drop out of society are those that can afford to.

The hippie movement was arguably California's first, more-or-less entirely homegrown socio-cultural phenomenon, resonating as it did with the state's youthful, experimental outlook. An outlook which often strained notions of propriety and tradition and which was quick to oppose the establishment. At the same time, it also reinforced California's reputation for superficiality and self-obsession; a cult of the individual.

Hippies and Silicon Valley may appear to have little in common, but the two did come together in both philosophy and individuals. Many of the Valley's achievements were driven by idealistic students and young people who were naturally also drawn to hippie culture. They came together with notions of forging one's own path and seeing the world differently from previous generations. There was also a shared sense of both the power of the individual but also of collaboration and community. There was a desire to change society for the better, and where some might see political, intellectual, cultural or spiritual routes to that change, Valley pioneers felt technology held the key to a brighter future. Both groups were unwilling to conform to traditional, conservative expectations and valued freethinking.

Although the hippie majority placed more emphasis on nature and a pre-capitalist world, they and the Valley agreed on the need for something to replace what had become a discredited system. A social and cultural order that had sown the seeds of war, corruption and division. Key to both hippie freedom and technological innovation is not being encumbered by old-fashioned ideas and looking in unexpected areas for inspiration. Even the internet, then still a somewhat unformed network, could be seen to have at its core hippy ideals. A dream of communal living, of equal contribution and taking according to one's needs goes some way to describing how early optimists saw the internet. Everyone connected, living together, with equal voices and an equal contribution.

As the 1970s faded from view and America struggled out of a socially, politically and financially traumatic decade, Silicon Valley continued to grow. Its economic power and influence steadily increased as more and more people and organisations relied on its products. The computer was a consumer product now, thanks largely to a thriving role in the home entertainment sector. Machines were comparatively expensive, so were sold to consumers as a way to both entertain, educate and run the administrative aspects of life – something for all the family; an investment in modernity.

Meanwhile more specialist computers and software meant applications in all areas of business from banking to creative industries. Machines that designed, accounted, communicated, recorded and analysed. Supercomputers, incredibly powerful (and large, complex and costly) machines were used in everything from weather forecasting (still vital to the world and its economy) to oil exploration. The Computer Revolution was well underway. The wealthiest businesses and governments all embraced computers as a way to improve their efficiencies or reduce their costs. Those late to the change invariably fell by the wayside.

By 1980 ARPANET was exchanging 100 million emails between 430,000 users, and other computer networks around the world were developing. In the same year, Tim Berners-Lee was working at the European Organisation for Nuclear Research (CERN) in Geneva. He proposed a new way of displaying text information that allowed the text to link to more related information instantly. With this idea he created a system called ENQUIRE. Ten years later the military stepped back from ARPANET opening the way for a publicly available computer network. Berners-Lee adapted ENQUIRE for the new network and the new computer technologies that were now more commonplace. This became the World Wide Web, a way of viewing information distributed over what was by now the internet.

With the opening up of this now global network, there came pressure for machines to handle more and more data. This in turn pushed the commercial motivation to create ever more powerful consumer computers as well as business machines. The typical home computer of the 1980s was ill-suited to the possibilities of the internet despite being sufficient for most people's more everyday needs.

By the 1990s the majority of homes, and pretty much every office in the West had a computer. The power of the supercomputers of the 1960s and 70s was now sat in most studies and student bedrooms. As such many people had the power to process, access and acquire large amounts of data, but most just did not have the need to. Although the market in business computing and the rise of the likes of Microsoft pushed the corporate world to trust ever more of its processes to ever more powerful, capacious computers, home users rarely needed such increases in capability.

One area that did see a domestic demand for ever-greater computing power was gaming. The main players of the time, Sega and Nintendo, later joined by Sony, were feeding a consumer appetite for ever better and faster graphics and larger gaming environments. The games console was actually the first computer device many homes embraced, with the likes of Atari's 2600 and the Nintendo Entertainment System. Even Fairchild attempted to break into the market, but would be largely overtaken by home computers. That market however was swinging back the other way. Consoles were now eating in to the domestic computer market and a burgeoning world of games programmers were creating a new sector that companies like HP, IBM and Apple were not really a part of.

The internet came to the aid of the big computer companies. The potential to access all types of content required computer hardware, not a games console. What started as a niche research tool and computer science resource found a multitude of commercial applications. Initially a home for news updates, hobbyists, and other 'virtual communities', a range of companies offering products and services over the internet began to appear. Alongside existing businesses offering their wares to internet users, internet-specific companies immerged. In the same year, 1995, as the first confirmed online shopping transaction took place, Amazon.com was launched.

These global shifts only increased Silicon Valley's influence and wealth. For 50 years the area had been a magnet for traditional technology companies. Businesses that created or improved software and hardware products or components gathered together to share people and ideas. Now there was an influx of a new type of company, one that offered services for and over the internet.

Moore's Law continued to apply beyond his originally predicted decade, and computer power at home and in commerce increased exponentially. Each year the potential of the internet to offer more, faster and superior content doubled. Each year computer hardware advanced, leaving the previous models in its wake. Each year hardware became cheaper. The conventions of business applied to Silicon Valley as much as anywhere else; rules of supply and demand, advertising, creating a consumer need, and product obsolescence.

Advances in hardware also meant the opportunity for users to create and share more content. In addition to being able to buy products and share text, they could increasingly share audio, photo and video material. Consumer demand now led developments, with ever-greater digitisation of content they wanted faster, cheaper and better hardware with which to create, access and share. Internet access left the preserve of computers and moved to a new generation of mobile phone as well as games consoles and other devices. Access to the internet, using it to share and communicate, to research, to shop, to inform and to entertain would become part of everyday life for most in the Western world. Even more than that, it connected most individuals in the West, and increasingly all over the world, to each other in a way never really imagined before.

Increasing connections, increasing content was not simply a supplier and consumer relationship; it became a two-way street. More users doing more things meant more data being consumed, but also more being produced. As users created content, browsed and bought products and services, it all offered a potential insight into who they were.

Where once businesses only really knew their customers via fairly imprecise surveys and marketing techniques; small samples used to represent a wider customer base, now there was a way of understanding every customer individually and accurately. Improvements in computing power enabled ever-greater amounts of customer data to be analysed and understood. It also meant more points at which to gather than data as people used computers to do more things. It was the dawn of the age of the Information or Data Revolution. It was a revolution that would change Silicon Valley and the world.

17

MEASURING HUMANS

After the Second World War, America grew increasingly concerned about Communism; not just spreading from the Soviet Union, but already within the US. Senator Joseph McCarthy was charged with rooting out 'subversives' within American society. Chief amongst his concerns, outside politics, were the influential worlds of the entertainment industry and academia. The former was centred in California and the history of the McCarthy witch-hunts in Hollywood is a long, well-documented and tawdry one.

In academia, the danger of young minds being corrupted was closely monitored. American paranoia was in part a response to the British 'Cambridge Five' spy scandal that saw well-connected Cambridge University students recruited to the Soviet cause and then going on to important government and establishment jobs. The American reaction was: root out the Reds in the universities and kill off the fifth column at its origin.

In the search for subversive influences many conservative

observers conflated atheism and Communism and focused their attention on university philosophy departments. The academic pursuit of philosophy was also home to other dangerous ideas. Ideas ready to challenge the belief that capitalism was the natural order of the modern world, even if it was a purely academic consideration.

The first Chancellor of the University of California, Los Angeles (UCLA) was Raymond Allen. He would go on to play a leading role in US psychological warfare strategy but prior to his time at UCLA he had come under pressure to dismiss professors that had joined the American Communist Party. To do so legally, he reasoned that any Marxist academic was obliged to follow Soviet ideas and propaganda and as such could not deliver the unbiased search for truth that education required. That meant the teachers involved could be discharged for not doing their job properly rather than for having a political opinion.

Allen, a philosophy professor, 'proved' that the political and philosophical bias of Marxist-Communism was unscientific - it failed to establish a hypothesis which could then be tested, and as such it was irrational, and therefore unbefitting a credible academic. There was no room for debate, interpretation or nuance in this; for Allen it was logical, irrefutable. This delivered a formula by which colleges could dismiss those declared to be, or suspected of being, Marxists or Communists. Furthermore, Allen spread this work across all the academic institutions of California, forcing job applicants to have their applications scrutinised by the state's un-American activities committee. It handed a veto over academic appointments to a non-academic body that no institution could fight, and it influenced education for generations.

Into this atmosphere was fed the work that the University of Chicago and the RAND Corporation did on Rational Choice Theory. People always making the right, self-interested choices meant free markets and democratic elections, neither of which operated in the Soviet world, was the correct, natural order. These two senses of the rational – Allen's scientific method and the rational nature of humanity – were drawn together to provide a guiding philosophy for America for the next 40 years.

The end of the Cold War meant that in the first half of the 1990s, the internet age met a new world order with the end Soviet Communism. A new technological era, a global era, was evolving at the same time as liberal, Western models of democracy and economics appeared to have indisputably won the day and taken over the world.

Political scientist Francis Fukuyama won over many observers with his somewhat unfairly reduced statement that 'history was over'. The ultimate triumph of the US-inspired free market, liberal democratic model meant there were no more major global social shifts to come. He, like many thinkers before him going back to the 1700s, felt that progress was leading the world ultimately towards one, stable state; a social and political equilibrium. At last, with the end of the Cold War, the world had gotten there.

This sense of one-world politics combined with the internet; a free, open platform for the exchange of ideas and information. To many, civilisation was on course to enter a new golden age. Conflicts on a truly global scale were petering out. Humanity could get on with making new things, finding new markets for the benefit of all. It would see the final success of globalisation. Just as innovations in shipping had effectively broken down physical geographical barriers to global trade, the internet would end cultural and linguistic borders. This utopia had been made possible by the incredible achievements of science; the summation of feats from Faraday to Alexander Graham Bell to Edison to Tim Berners-Lee. Science and rationality had ended history, and with it ideological and religious conflicts.

Fukuyama revised his thesis in the wake of 9/11, but for a decade the idea that a sort of cultural and historical equilibrium had been reached found favour. We now look back on the idea that history could possibly have 'ended' with a single geo-political, ideological shift as naïvely hopeful and somewhat obtuse. What it did mark was the end of a sort of post-war certainty. The Cold War cast its shadow over so much that social and technological predictions largely played second fiddle to the ideological battle. Now, aided by the internet, everyone would find a voice and the world would know just how fragmented and disparate it really was.

In those optimistic days of the mid-1990s, computer science students at America's leading universities were submerged in a world of internet possibilities. One such computer scientist was quick on to the field – Jeff Bezos set a template for internet retail by allowing users to find almost any book, cheaper and quicker than in a physical store. Having already worked on Wall Street for ten years, he helped build international trading networks and saw an opportunity to cut out the middlemen (retailers) from the author/publisher/reader chain. This disintermediation became a model for many businesses to come and it allowed Bezos to get in early and start to dominate the early internet shopping sector.

A couple of years later, Stanford students Larry Page and Sergey Brin were looking at how people searched the internet for the pages they needed. They developed an algorithm designed to rank the suitability of a webpage rather than just its popularity (as other early search engines did). They refined this algorithm, called it Google, and launched it to the world in 1998 (and like Hewlett-Packard and Apple before them, briefly ran the company from a garage in Silicon Valley).

In the Valley and in Wall Street, a wave of companies were striking out, predicting a potential gold mine in the internet. The chance to offer services and products to millions around the world was too great an opportunity to miss. From the mid to late 1990s scores of companies were founded, usually by energetic students and graduates, many with extensive technical knowledge but often little business grounding. Investors, conversely, failed to fully understand the businesses, the people behind them and the potential markets. A market bubble immerged as money poured in from investors believing that the profit from those companies that worked would offset the losses from those that failed. Share prices in these companies kept rising as long as people believed in this new breed of company.

A combination of poor stewardship, short-termism and outside influences from 9/11 to a global stock market downturn saw dozens of early internet companies go bust. Some, such as Amazon and eBay, kept their heads above water and weathered the storm. Those that survived emerged into a new internet age. The capacity and speed of computers and telephony continued to increase, as did consumer penetration to the point where around 50% of homes in the West had internet access. Dubbed Web 2.0, this environment saw the creation of an internet which was populated predominantly not by content produced and controlled by businesses, but by content produced by users.

Ordinary people were uploading not just text but photos, audio, and video.

In 2003, Mark Zuckerberg, a psychology and computer science student, headed a (not entirely collegiate) group that developed an online social directory of students at his university, Harvard. Within 18 months Facebook had moved to Silicon Valley, had millions of dollars invested in it, and was on its way to connecting people, and their content, around the world. For a few years Facebook looked similar to many of those late-90s companies that offered something attractive to users, but had no way of making money from them. A business that would grow rapidly, burn through cash, and hope it could be sold or a way to make a profit would become apparent before the money ran out. The difference between Facebook and those earlier, less fortunate companies was that now it was users, not those behind the business, that kept it going. It was users providing content, activity and data. It was users that would become Facebook's most valuable asset, and the asset that would repay investors.

The user-generated content phenomenon radically changed many Silicon Valley business models. Hardware had to provide more and better means for users to generate that content, from devices including cameras and microphones and a range of new input devices to new forums and platforms to write and share ideas. For those providing internet services, it meant huge opportunities to learn ever more about their customers.

The user-generated internet also suggested something else; that America had been right over the last 50 years. That this new era was about the power of the individual; the worth of the individual and their capacity to create their own wealth and satisfaction. Human Capital as described by those influential University of Chicago professors and that merged into the Cold War ethos of the US really did win out. Individuals would now be allowed, even encouraged to create their own worth on the level playing field of the internet. All you needed was a computer and a connection and an idea you believed in and you could change the world.

The American Dream. The socio-political and economic primacy of the individual and their freedom. The ultimate victory of rationalism and choice. Progress that is synonymous with technology and science. These were the beliefs that the twenty-and-thirtysomethings of Silicon Valley were raised on and believed in unquestioningly. However, like many of their forebears, particularly those of the 1960s and 70s, they entered the tech sector seeking to improve the world and prove their theories. They were driven idealists, convinced of their personal right to succeed but also to pioneer; to discover new ideas and new ways to do things. To find new applications for technology.

The internet was a vast, hopeful new world. A tool to bring the world together and allow anyone access to any knowledge or experience. It would make life simpler for everyone. Google would mean you could find out anything you needed to know or access any service you wanted. Facebook would break down barriers of geography and allow new voices and ideas to be heard on a scale never previously dreamt of. Such lofty ambitions, however, required funding. No matter how noble the aims of the founders; no matter how novel these businesses were, they were businesses subject to the same pressures of capitalism as had existed for over 300 years.

As these companies and their peers soared in influence

around the world, they recruited more experts, developed and implemented new ideas, and built ever more elaborate infrastructures. Money was invested, shares were sold, and so, inevitably, profits had to be made. In the case of many of these companies, entirely new in their nature as they did not actually produce any physical product or provide a conventional service, their business models were simplistic. Most had no direct source of revenue, relying entirely on a constant source of outside investment. A few went with a subscription model of some sort. Many saw advertising as the best route to stability.

What investors saw, and in time tech companies came to embrace, was the opportunity to gather untold levels and types of personal data from users, and the potential value of that data. In their use of social media, search and online retail, and later media consumption, an unparalleled picture of who an individual was could be built up. And people would pay for that level of knowledge; that level of insight into groups and individuals.

THE SOURCE OF ALL DATA

"I will not be pushed, filed, stamped, indexed, briefed, debriefed, or numbered! My life is my own!" "I am not a number. I am a person." – **'Number 6'**, ***The Prisoner***

Opinion polls, it is generally agreed, were born in Pennsylvania in 1824. One of the oldest American states and home of the Declaration of Independence. This keystone of modern democracy was the testing ground for a new idea: asking people how they intended to vote before they voted.

It was the Presidential election fought between the controversial populist Andrew Jackson and the well-connected establishment figure John Quincy Adams. It would prove to be a tense campaign with four candidates all of whom essentially represented the same party. Only Jackson came from outside the political elite. He was a hugely popular general in the US Army but he had to be persuaded to run, in part, to make a stand against the perceived corruption of the incumbent administration.

The State of Pennsylvania had also led the way in enfranchising more of the population. By 1824 most of the United States allowed almost all white males to vote - previously voting was restricted to white, property-owning males in common with most of the Western world. With an electorate now spanning a much wider range of socio-economic backgrounds, Jackson, a man of the people who had led American forces to victory over the British at New Orleans, was increasingly looking like the favourite.

The political establishment was unsettled by the potential for a shock result. Jackson's popularity saw his opponents refuse to criticise him in public, but in private they planned to unify their support and, headed by Adams, defeat him. It was in this atmosphere that political groups started to use straw polls at town meetings to test which way the popular wind was blowing. In Pennsylvania, The Aru Pennsylvanian newspaper became the first to publish the results of such an unofficial poll.

The poll suggested a comfortable Jackson win in the state, as did the other straw polls, pushing Adams et al into a manoeuvre that would poison American politics for the next four years. Jackson, in line with the polls, won the popular vote, but with the support of the other candidates Adams took the presidency. Whilst Jackson and his supporters were outraged, laying the foundations for a bitter, personal rematch in 1828, the power of the polls had been confirmed. Asking a representative sample of people, extrapolating that number to account for the whole population, and looking at the result could accurately predict elections.

Despite this early indication of predictive reliability such polls generally remained localised until 1916 when The Literary Digest posted millions of postcards to households in an attempt to gauge national opinion. This method correctly predicted the following four Presidential elections. On the fifth attempt (Landon v Roosevelt), it failed, although seminal pollster and advertising researcher George Gallup called it correctly.

Politics was, inevitably, followed by business where public opinion, although often gauged by success and failure in the market place, could be a vital influence in decision-making. It also had the potential to help avoid costly mistakes. As more people wanted an insight into the popular mind, opinion polling and its commercial cousin market research became big business by the 1930s and 40s.

The obvious flaws of polling and research - you could not physically ask everyone, many people did not really know what they wanted - did not seem to stop the powerful and wealthy from seeking its insights. Henry Durant, the pioneering British pollster, was remarkably honest in assessing the profession he did a great deal to establish. He suggested it was the stupidest of professions, for why else would someone publish a prediction on Thursday morning that may be proved wrong on Thursday evening?

Despite this, the insights offered by asking the public started to be studied seriously. Methods of how to get a more accurate picture, with more robust results were explored. Where choices were strictly defined, as in an election, the task was relatively easy – ask people for whom they are intending to vote, and chances are, by election day, they will not change. Compile results and surveys over numerous elections and it will even become evident the number who are likely to change minds. Assuming you can ensure a representative sample of respondents, you will have your prediction. As such opinion polls tended to stick with binary questions of a political nature – who would you vote for; should the country do this; would you support such a decision?

In business, Gallup and rival pollster Daniel Starch both worked extensively in early market research along similar lines. By asking interviewees whether they remembered certain adverts or messages, and compiling the responses, they provided a powerful new tool to advertisers. No longer would it be necessary for business leaders to simply use their best guess on what resonated with their intended audience; they could employ a company to go out and ask the public whether it worked or not.

Away from political campaigning and advertising, public surveys had a more serious role to play. In post-war Germany polling was used monitor the reduction in popular Nazi sympathies throughout the country. After over a decade of indoctrination and propaganda, it was necessary to measure personal and popular sentiment. There was, however, so much data regarding former party members alone that IBM were asked to provide a machine to help with the analysis. This would herald the advent of computerised polling, enabling the processing of larger samples and more complex questions.

Whilst originally there was little or no consideration

for who was responding to a survey or poll, beyond some basic attempts to incorporate geography and wealth, the more money that became involved and the more decisions they affected, the more accurate polls had to be. That meant better accounting for the variety of the population and their ideas. It was impossible to ask every single voter for their voting intention, or every householder for their choice of soap powder. So a representative sample would be contacted. This required an understanding of who could accurately represent a group, and how big that group was as an overall proportion of those involved in the final vote. That involved breaking down a population by age and gender, of course, but also economically, geographically, educationally. Once segmented and labelled, the population could be divided up for a wide range of questions, not just elections.

In the commercial world companies would come to rely on research to confirm more than just press or radio adverts but everything from the quality, or desirability, of a new product to corporate logos to pricing policy. For decades those leading businesses had been assessed, in part, by their ability to understand their market and to meet the needs and wants of their customers. Now data would be used to replace, influence or support their decisions. The marketing manager or even chief executive could ask for research instead of following experience, wisdom and intuition.

It soon became clear that the reliability of a survey could be significantly affected by the wording of the question asked. 'Do you find politician A more trustworthy that politician B?' implies that the respondent trusts one of them, when in fact they may both be as shifty each other. When asking multiple questions, one question can influence the next. How the question is framed also influences respondents, such as 'Should those concerned about the health service trust politician A?'

This sometimes quite accidental bias in surveys added another layer of complexity to an already complex process. Questions needed to be carefully worded and entirely objective to deliver robust results. Positive or negative adjectives should have no place in a question. Nor should the appearance of assumption.

That is, unless you want a particular result from your poll. Something that gives the appearance of evidence for a decision made. A result that motivates a particular group into action. It is not unknown for those commissioning a poll to do so simply to prove their hypothesis right. The survey, lacking any other substantial evidence, would be used to counter arguments from others. Equally, whilst those commissioning the research might not say they want a particular result, those carrying it out may feel, in order to be commissioned again in the future, they should deliver something that confirms a particular strategy or opinion.

As if this was not complex, and by extension unreliable enough, polling also required an ever more intimate knowledge of the respondents. In order that they could accurately represent the mass of the population they must be broken down into representative groups. Income, gender, education, geography were all key, but potentially so were tastes, political opinions, and all manner of psychological traits deemed potentially useful in knowing where biases might lie.

Polling moved from face-to-face to the phone and then the computer, but each process held inherent biases. What time of day were you on the high street asking people? Who is walking around at Saturday or lunchtime on a weekday? Who answers the phone at 4pm on a Tuesday? Who has access to a computer and email? All of these have changed over the years and all can mean a very different type of sample is being surveyed. Even those that have said in the past they are willing to participate in a future survey means they already display certain attributes. It would be almost impossible to correct or account for many of these problems - although some researchers would try.

Commercial and political polling and research divided people socio-economically and psychologically with reductive categories. People were labelled AB if they were managers or professionals, C2 if they were skilled workers. Personality tests reduced people to introverts or 'analyticals'. Where someone lived, how they spent, the car the owned, their weekend activities all started to form a basic picture of who made up a population. Each characteristic assigned numbers in order for them to be analysed and accounted for en masse.

This reductive way of working, of being able to score and measure numerically, played comfortably with the digital age. Computing was already used to compile polling numbers – simple yes and no responses. In the digital era companies could combine their data with others. The data they had about who bought their product could now be merged with other, publicly available data, and with data bought from specialist organisations. In the early days of market research, companies would look at census information to understand the relative wealth of certain parts of a town. Now everything from whether they owned a pet (and what species) to where they holidayed to their favourite drink was being recorded.

Added to conventional research data acquired through direct contact with customers came methods that overcame the conventional barriers to understanding people. Generally, you had to offer people some sort of incentive, usually financial, and that skewed research. They would answer differently, and certain people would be more likely to sign up. But if you can take their data without them really feeling it, without it inconveniencing them, that would be much more valuable.

One of the groundbreaking methods in consumer data gathering came in the shape of the loyalty card. Knowing who customers were and what they bought changed retail operations. By monitoring what people bought predictive models enabled retailers to more accurately buy in stock, which in turn made for more efficient supply chains, smaller warehouses, and less waste.

As universally helpful as that might be, just as valuable to companies was the ability to build up a picture of who customers were. There was no need for them to tell you their personal circumstances directly when you can read so much into their weekly shop. Whether they have children or live alone, whether they take their health seriously, if they have a cold or, as shops diversified their offering, their hobbies. This became and remains a hugely powerful tool by which retailers can customise marketing messages.

Loyalty card and similar data would soon be overtaken in size and scope by online data where even more behaviours and attributes could be measured, labelled and scored. Applying psychological and behavioural analysis meant even more could be assumed about a person. Increasingly detailed pictures and, in principle at least, ever more accurate predictions about who they were, what they wanted, how they saw the world and who they would vote for. You no longer had to directly ask them.

A big issue, in politics in particular, is the effect of the poll itself. Even if people did not know they were being polled, the outcome still potentially affected public sentiment and, paradoxically, making any prediction suddenly less likely. If people believed that the election would be a landslide for one party, why bother voting. Parties want to look like inevitable winners, overwhelmingly the better option and supported by the public because the people are wise. As voting day nears that certainty in the minds of supporters might become complacency. This can affect the desired result of a poll that a party (or its media or corporate supporters) might prefer and as such, if they are commissioning a poll, they may choose very particular wording. The results may also be released at specific times during the campaign.

Time and again, sometimes in business but certainly in politics, people seem to almost deliberately avoid attempts at being predicted. Especially when those doing the predicting were seen as part of an elite, many of whom the public had come to distrust and against whom they reacted. More recent political polls have witnessed the phenomenon of the secret or shy voter; the voter who, sensing it was perhaps thought socially unacceptable to vote a particular way said they would vote the other. Or that giving a certain answer would lead the questioner to draw certain conclusions about them that they were not comfortable with, so they would deliver a misleading answer to pollsters.

The rise of social media, as both a source of opinions and an influence on opinions, has added a new element of unpredictability to surveys. Pollsters have enthusiastically adopted social media as a means of both accessing a very large sample group and also understanding much more about the individuals questioned. Social media users are not necessarily typically representative of a wider population – for instance they tend to be younger and more urbanised for a start. That aside, social media has also become an unreliable source of news and information making it very difficult to know what stories are making an impact on voting intentions. Rumours and lies are broadcast, assumed to be fact, unchecked, and used as a basis for further comment, purchases and decisions. Social media profiles and their news feeds work on Confirmation Bias. They cause information to be filtered, under the guise of removing that which is of no interest to the user, so that no conflicting opinions or contradictory news is displayed.

One major problem with democratic politics is that it has to renew its obsession with public opinion every few years and the public are quite inconsistent. Nonetheless, those in charge must try to understand them, their opinions, and what it will take to be elected. Polls are supposed to aid that understanding, but modern humans often hold contradictory, inconsistent views. They change at a whim (or at a headline), they ask for things not in their own best interest. They pay attention to one fact, but ignore another.

With all of these biases it is unsurprising that no matter how much data there is on people they remain, on large, unpredictable. No matter how carefully selected a polling sample and complex the formulae to account for the biases they and their lives are just too complex. Well-publicised failures and the ever-increasing complexity of influences have not put the pollsters off. Although it has dented confidence in them amongst the public, the commercial and political establishment retain faith with them, perhaps because of a lack of any alternative.

Like economists, pollsters have relied on rational, predictable behaviours and decision-making combined with probability and margin of error. They have assumed that maths and computation would analyse reliable data and produce accurate patterns. Like economists, the pollsters' art is not completely undermined. It provides genuinely useful insights, but over-reliance needs to be guarded against. Their results need to be placed in context and viewed with scepticism. They should equally not be a scapegoat for poor planning and decision-making.

Also like the economist, the pollster believed that despite huge volumes of data being provided it was still a lack of data that was the main barrier to accurate prediction. So more data was demanded, and the online world, and the ever-increasing amounts of life being led there, readily provided it.

The thirst for data, and the online well that attempted but never succeeded in sating it, created another issue; who owned the data? The person who generated it or the company that gathered it? The person measured had done almost no work whilst the company had developed systems, employed analysts and provided a service. Data - weightless, endlessly replicable, and worthless without context or processing – demanded a new economics.

18

DATA HAS ITS PRICE

From 2005, more and more internet services came to rely on people creating content and sharing information about themselves. What they did, thought, enjoyed and disliked; it all became the lifeblood of a new internet era. Facebook was joined by Twitter, Snapchat, WhatsApp and Instagram – companies that made nothing, but offered a digital platform for messages and content to be exchanged and consumed.

True, in many ways, to the founders' visions, social media and the technology that enabled it was largely egalitarian - classless and borderless. Anyone could and did become involved. The companies and those that funded them found new ways to appeal to every group and market. In many cases their products started out as a useful tool in business or life. Those products then started to take over every other aspect of what they did, consuming leisure time and becoming a vital part of users' lives. They claimed to offer freedom and insight, to be an instrument to improve everyone and everything. It could save time, make money, tell users about themselves, tell them about others, and offer vital advice, news and information.

Google came to dominate search and, although it failed (by conventional business standards) in its attempts to join the social media sphere, it bought YouTube which became the biggest video platform and the second biggest search engine. Amazon grew to become the world's largest retailer selling almost everything to anyone anywhere. Apple, although not inventors of the smartphone, popularised the device that did not just enable internet access from anywhere, it also generated yet more data on the user, in particular, location. In turn, its operating system meant that anyone designing an app to use on its devices had to work within their parameters, a model others went on to emulate.

Both Google and Apple provided the platforms and operating systems for apps that meant they in turn gathered all the data used by those apps. Data on exercise, travel, social interactions, news of all types, media consumption; all of it gathered by those behind the software and the platform it used. Facebook also moved in, connecting apps to social media profiles and drawing together an ever-wider assortment of data.

Every interaction, every photo liked or message forwarded became a small part of a greater psychological profile. Each element another pixel in the high definition picture of who the user is. Tech companies took digital trails of searches, likes and views and dumped them into vast silos to be sifted, labelled and matched, revealing more about users than they knew about themselves.

The companies that now owned and protected this data found a ready and lucrative market for, if not the data itself, what it revealed. The trends, insights and predictive patterns that could be found within the data. Advertisers could now see and access every customer on an individual and group basis; their behaviour and their personality traits. The sheer amount and variety of information being offered up enabled complex, detailed pictures to be established. Psychological and commercial profiles which allowed advertisers to both design their ads and target them most effectively.

Social media enabled businesses to know where people were, when, what they did; their names, their partners' names, their favourite film, who was invited to their birthday party and at which restaurant. Added to that was what they were searching for, what they clicked on, and what they bought as more everyday tasks were carried out online, delivering ever-greater insights.

As if that was not enough, apps, games, tests and surveys were specifically and carefully designed to reveal commercially useful user details. Psychology had long used tests designed to reveal information about individuals and to build up pictures of particular groups. Now these tests were rebranded as quizzes, given a cutesy or practical spin and used to reduce people to 200-300 specific, commercially applicable data points. They were often under the guise of enabling the user to learn something about themselves or the world. Inadvertently they revealed their way of thinking, their opinions, biases and preferences. These games and quizzes, health monitors and time-killers became a part of users' lives and their leisure time. The makers of these assorted tools harvested the data users submitted, but so did the platforms they worked on. Facebook, with its almost sole reliance on advertising income made particular use of this vast new asset it acquired.

Data – getting it, storing it, interpreting it, selling it – fuelled a massive expansion of Silicon Valley, as well as a shift in its goals. It was the second gold rush as thousands of entrepreneurs rushed to establish companies all providing some service that would persuade people to part with a bit more data about themselves. Like the original California Gold Rush, where rules were, at best, vague, the data was out there, waiting for someone to stake a claim and take it.

Unlike gold, however, data is not a physical commodity. It does not have an agreed upon value that could be traded independently. It cannot be weighed, scientifically assessed for its purity, and exchanged for other physical goods. It is not a currency backed by gold or an obligation backed by some other asset. It is perhaps for this reason that users saw no real value in their data. People readily gave up what was not valuable to them in return for access to an incredible array of services from free videos to email, the chance to connect with people, to broadcast one's opinions, to feel important, and to be liked. Further, they were not giving away their data; they did not actually lose it to someone else, it remained theirs to give away again and again.

Users may not feel their data has much real value, indeed, they might even feel their data is a commodity they are happy to part with in exchange for the services of Facebook, Google and other. Their data, however, does have a great deal of value to businesses. Collecting and collating this data gave the tech companies the opportunity to sell unprecedented insights into consumers so that they in turn could be sold to.

Another unusual characteristic of personal user data is that it has relatively little value to advertisers on its own. Data is a uniquely granular type of commodity. Gold tends to have a value no matter how small a piece you have. Data only really gains value when there is a lot of it. When it represents a lot of different aspects of a lot of people and trends and patterns can be derived.

Unless you are incredibly important or influential, almost no company would pay much for a psycho-socio-economic profile of you. No one company cares a great deal about what you spend your money on, whether you like cats or dogs or the colour purple, who your social media 'friends' are, or whether you are characteristically analytical or emotive. They do care, however, about thousands of people in a country or market that might all want to buy house insurance, when they are likely to buy it, and whether they are likely to respond to an anthropomorphised dog telling them to do so.

Data only has value once collected and filtered and an application found. A company able to say they can specifically target certain potential markets with certain well-designed messages is tempting to business. As such, it can be very rewarding for those companies holding the data – and that design the algorithms to find patterns within it. Especially in an age where the process from advert to purchase is usually just a couple of clicks.

As long as marketing and advertising has existed it has wanted to understand the target consumer better. It is said that 50% of all marketing is wasted - that is, half of those that are exposed to it are not going to be interested - but no one knows what 50%. Technology and data allows advertisers to reduce that waste by understanding more about a person, their habits, their interests, their spending, what they are currently looking for reviews on, and the types of messages that might resonate. This increased market segmentation and made it more precise. It made the targeting of messages more accurate. The value advertisers placed on these advantages has provided the bulk of the revenue for many tech companies.

These insights also provided the opportunity to communicate with consumers, to refine products according to these new insights in a way that previously would have been limited to expensive, imprecise, small-scale, in-person surveys and focus groups.

In Richard Serra and Carlota Fay Schoolman's 1973 short film *Television Delivers People*, they suggested that, contrary to appearances, television networks supplied viewers to advertisers. The advertiser, the company selling their product to the public, is paying the broadcaster to deliver the public to them. The successful broadcaster will make programmes that hold the attention not necessarily of the most people, but of the right people; those that spend in the right way as determined by the advertiser.

Serra and Schoolman go on to criticise the intellectual level of programmes, of the content supplied by the content providers; the conspiracy between broadcasters and advertisers, and the nature and type of culture that television is responsible for. At base, however, is the conclusion that if you are not paying for something (as, for example, in the way US network television is not paid for directly by the public) it is you, the viewer, which is the product being sold (in this case, sold by the broadcasters to the advertisers).

The idea that a human (or their data) is the product being sold has gained renewed traction in the internet age. Where broadcasters produce programmes that are an educated guess at what certain groups want and will respond to, and therefore access to whom certain advertisers will pay for, the internet can target much more specifically. It can also prove its impact a lot more convincingly as each view, click and second spent watching or reading can be monitored.

Now the shared mission of Silicon Valley was to find new ways to get more people to part with more personal data. To get them to reveal details that may have value now or at some future point, especially when combined with other data. Data science, the use and analysis of data, the discovery of patterns and of making of predictions, became the dominant Valley activity. Advances in technology were now virtually on a par with mathematical and psychological research. Thousands of PhD graduates were employed to find new ways to get users to spend more time giving away more data, and to find new ways to use that data.

Foremost amongst the tech companies are Facebook, Google and Amazon who harvest data, and Microsoft, Samsung and Apple who make much of the means to enable the harvesting. These are unquestionably the biggest, most prevalent, all encompassing companies in the world, not just in the digital domain. One of the main reasons for that is their unparalleled ability to understand humans through data.

Like US network television, Facebook and Twitter and Instagram and dozens of other, smaller services are monetarily free (or very cheap) to use. These are the platforms for millions of messages, preferences, feedback, ideas and interactions that, brought together, build up detailed pictures of users. Google is both free and arguably the most indispensable service on the internet (certainly within most of the Western world). It is the default for hundreds of millions, enabling them to find news, shops, recommendations, reviews. It is the default because most of the time it reliably sifts through vast numbers of webpages to find what people are looking for, or something close to it. They are willing to take on trust that it does so in an honest, unbiased manner. The ability to do this does not come cheap.

In return for the free searching and content, the free connections and distractions and information, these services accumulate data. They know what users are looking for, how, from where and when. From that information, data engineering connects and extrapolates, building up a picture of each person, gathering them together under certain categories, the better to target them with adverts at just the right time. Data has become the commodity and the currency of a new trade. Bought and sold, the value of data can change, it needs to be (to some measure) securely stored and protected; people even steal and covet it.

Every few years a new company joins the Silicon Valley big players. Netflix do not advertise in the way Facebook does. Their model involves building up a picture of viewers to better create programmes that they will watch. They understand the psychology of the viewer and apply this to narrative structures and characters in their films and programmes. Their business is built on predicting what viewers will want. The Netflix service, like Amazon, is not free, but that does not mean they do not see value in the data that is available to them. Indeed, they use that value as a way of delivering services at a minimum cost, thereby forcing competitors out of their markets.

What started as a way of reducing the amount of wasted advertising by understanding more about the disposable income or age of a user has become more and more carefully targeted. The, if not benign, then at least not-entirely-malign goals of Brin and Page (to open up the internet to all), Zuckerberg (to connect the world), Bezos (to remove profiteering retailers), Musk (to remove profiteering banks) and others had, eventually, given way to the iron laws of capitalist economics. They must keep innovating, growing, increasing their assets and making more money.

We now search for everything via Google - what to eat, where to holiday, what the best toothpaste is, what a celebrity is wearing, what car is the best performing, the latest film trailers, and where to see comedy on Saturday night. Many products are bought through Amazon, and experiences and opinions are shared on Facebook. We pay, without too much thought, through PayPal. The GPS in our phones (usually Apple or Google via Android and Samsung) tells companies where we are, where we live, our route to work, and for how long we are there. These phones with their apps that help us, entertain us, inform us, challenge us, make tasks easier, pay for purchases, monitor our progress, enable us to work from anywhere all feed terabytes of data about us to tech companies.

All of this data, derived from careful psychological examination, draws a picture of who we are, what we do, and, being creatures of habit, what we are going to do. Ads are no longer just ads but a tempting, strategically placed and timed portal into the shop where a purchase is so frictionless as to be made instinctively by our System 1 brain.

Amazon Prime allows users to purchase just by asking an Alexa device to buy it. Payment hardly comes into it. Uber allows users to be found by a taxi, taken to a destination, and then pay with little more than two presses of their phone screen. A future of driverless cars means we will not driving so can be sold to and monitored more. Home assistants like Alexa or Google Home, once a science fiction fantasy, listen to us, waiting to be instructed to turn on lights or the heating or to buy some snacks or change TV channel and is seen as the ultimate data gathering tool.

Amongst the myriad of sectors Google are moving into

is mobile telecommunications. Users of the company's Pixel phone, and in time any user of their Android platform, will, rather than use a service provider like Vodafone, Telefónica or Verizon, just use Google. The system relies on an eSIM, a non-physical SIM that allows users to connect to any network based on what service is providing the best signal at the time. Users receive a seamless service wherever they are, and with an increasing range of devices, and at a significant financial saving (just under 50% of current costs). Google can do this by using their size and financial strength to force infrastructure owners to offer them cheaper deals. Google can furthermore afford to make almost no profit on the service because a user and their data are far more valuable to them than they are to a traditional mobile telecoms provider.

How did Silicon Valley get to this point? Where the manipulation of human behaviour, based on the large-scale, skilful acquisition of data, is so central? That it made a lot of money fast is one factor. It out-paced government - on regulation and tax amongst other concerns - and, to an extent, even came to dominate government. It embraced new ideas, not just of the scientific type, but of the social and psychological type. Like Richard Gatling, Alfred Nobel and Robert Oppenheimer, the Valley's pioneers did not always realise the potential negative power of what they had created. Once it was created and unleashed, however, the Valley could do little to stop its momentum. Science created the tools; tools that were non-political, non-moral, non-judgemental. Capitalism, politics and economics with its worldviews and vested goals and all-too-human ideologies used those tools. Science and nature clashed with the unnatural, the man-made. In much the same way, neuroscience, the science of the brain, would be drawn into technology and would also be applied in a way its pioneers did not necessarily foresee.

THE CULTURE OF TECHNOLOGY

Marcel Duchamp, the father of conceptual art, established his credentials in what most would consider conventional art. That done, he then set about bestowing the title of 'art' upon whatever object he saw fit. In doing so he challenged the idea of what art was, provoking a debate, and at the same time making the idea, rather than the execution, central to what art was.

From that point art became increasingly elitist and exclusive; an intellectual exercise that fetishised novelty. Critics and buyers, weary of seeing the same old thing, fell in the love with the new just because it was new. The daring, the experimental was now also the great. The cult of the new was a badge of intellect, a demonstration of one's insight. As such the new became as important to business and government as it was to art. How to find the new is now vital, and the process of creative thinking, of innovation, is an industry in itself. Organisations pay millions to individuals and companies who can help them be creative.

'Think Different' was the ad campaign and mantra of Apple when Steve Jobs reasserted control of the then-struggling business in 1997. Part plea, part command, the company anchored itself to innovation from that point. Not just innovation within, but without as well. With their product, we, the lowly consumer, could unlock our creativity. We could express our unique insights, our own new ideas.

Jobs subverted received business wisdom, and has been

lauded for doing so. He commodified creativity whilst also viewing people, his customers, as imperfect; the flaw in his perfect machines. These machines would make the imperfect human perfect. He bludgeoned his way through the public consciousness. He told customers what they wanted and never listened to them. For years Apple were something of an industry joke; incompatible, niche, over-priced, over-simplified, and hard to fix. Jobs' bloody-mindedness, however, won through. People fitted into his vision because they were seduced by his passion for perfection. Apple created products that claimed to foster the individual and their expressive, creative nature. They yearned to believe they were unique; not a dull, predictable drone of industrialisation. They wanted to be Apple, not a PC. They wanted to be a range of carefully selected colours and modern curved designs, not a beige box. They wanted to be Jobs. He was a role model for individualism. He has joined the ranks of those rebels and originators he aligned Apple with in that Think Differently campaign.

Everyone wants to be right. Everyone wants to be the rock star of their office, of their industry. Everyone wants to be feted, recognised, and heard. Chilean economist Artur Manfred Max-Neef stated that the human condition meant a need for subsistence, protection, affection, understanding, participation, leisure, creation, identity, and freedom. Silicon Valley would set about providing as much of these basic needs as possible, in some cases redefining them in order that they might provide them better.

Jobs was a man who, at least to a large extent, came to believe the myth he had created for himself. The claim that he was a genius is questionable. If his particular brand of insight, determination, obsession and flawed personality were indeed the marks of a genius, it was a particular and peculiarly modern type of genius. Unlike others, Jobs' genius was not one that could easily translate to other times, other conditions. It was uniquely applicable to his time, place and circumstances, but it made him into a Silicon Valley colossus. Regardless, of his place in history, Jobs does bequeath one particular lesson of dubious merit – that it is possible to persuade millions of people to part with money, time and personal information in return for something that whilst flawed, made them feel a part of something unique and new.

A constant flow of novelty and innovation also reassured investors. Share prices were buoyed by bold visions for the future. Investors frequently did not fully understand the industry they had invested in, so the appearance of progress had to suffice. Silicon Valley is often seen as falling prey to the novelty fallacy - the assumption that the new must be an improvement on what has gone before, regardless of what others think. In this, technology companies are often seen as being out of step with the wider public who tend to be more conservative. But being an industry built on innovation, tech companies pushed on with attempts to convince the world that everything new is positive. In regularly and repeatedly having the new forced upon them, however, the people increasingly came to feel overwhelmed, insecure and resentful.

Silicon Valley, by the 2010s home to the world's biggest companies, had settled into a pattern that had two distinct, but interlocked branches. One, naturally, was its foundation of pioneering technological progress - a cycle of ever improving hardware to facilitate ever more complex applications running ever more aspects of people's lives. The other was the cultivation of what Silicon Valley stood for; its ideology.

The Valley became shorthand for a culture; a way of thinking and of viewing the world and its problems. There came to be a Silicon Valley philosophy that affected a way of working and living. It involved giving over large parts of your life to far more efficient and effective technologies. To hive off the frustrating bits of life (some of which technology had inadvertently created anyway). In doing so you could focus on the productive and the personal. It was about the individual. The individual's ability to achieve, to develop, to be at the centre of everything.

These two philosophies crossed over in the generation of new ideas. The more time you had to think, to 'be', the more new ideas you could have, the more time to experiment, to fail and learn from your failure; to relax and be open. In many ways, the hippies' ideals never left this part of the world. They just took on a sleek, modern appearance and became much better acquainted with science and technology and capitalism.

Understanding yourself, and by extension others (be they employees, friends or customers) became a key goal of Silicon Valley. The Valley's business practices started to define modern business more widely. Google's much vaunted (but somewhat erroneously understood) encouragement of employees to undertake their own, personal projects. Netflix and their radically pared down employment and HR contracts. Amazon's scale and team structure. Facebook's 'fun at work' culture. Around the world companies from the newest and smallest to the largest with centuries of history wanted to understand what fuelled Silicon Valley growth. They were desperate to understand and emulate their mindset of innovation and success without, so it seemed, much of a downside.

This understanding, whether it is how to attract new

customers or get the best from your workforce, provided a ready home for behavioural economics, behavioural psychology and neuroscience. These developments could now leave academia and healthcare and make money. Furthermore, those researching these subjects could further their insights through the access to users that technology afforded. Huge psychological data sets were available and instead of asking a few dozen people in person, researchers could design apps to profile millions.

The focus of Silicon Valley had always been about technology, not how to attract, retain and exploit users. So it looked to others for more ideas-driven ways to organise data and extract more meaning, more applications, and more predictions from it. New ideas generally, not just innovations in hardware and software, would be greedily consumed by tech companies.

From new definitions of resilience to the importance of sleep. Processes to make you more productive or more mindful (mindfulness being Silicon Valley's version of hippie philosophies and practices for the connected age). Modern concepts of leadership, creative thinking, collaboration and everything from luck to boredom. There flowed a desire for more and newer perspectives on everyday, universal subjects. Like the well-worn marketing technique, concerns and neuroses were effectively invented, studied and packaged. Worries that people never previously knew they had were suddenly serious flaws that needed to be fixed. Luckily, technology was there to help.

Many businesses in the Valley had been founded with little idea of their long-term goals. By introducing a service to market the idea was that users would decide how they wanted to use it. With enough time and funding, and possibly a bit of carefully managed failure along the way, the user would guide the company towards a profitable application far more effectively than the company founders. Then the business would 'pivot' and find a clear direction and a way to make money. In the same way ideas were embraced, tested and applied to different technologies, and hopefully some would produce an income – that income could be money, or if could be data.

The need for ideas to fuel Silicon Valley growth saw a new market in 'thought leaders'; authors and public speakers who found new perspectives on old ideas. Perspectives presented in such a way that they did not really challenge accepted business philosophies. Instead they basically agreed with the new tech elite; empathised with them and provided an intellectual proof that they had the right answers. Just like economists who, given a political theory, could find the economic evidence to prove it right, now business practices found wider concepts of human nature to prove they were doing the right thing.

The work of thought leaders was scrutinised for commercial applications. As a result, for example, sleep becomes something to be bought and sold online. Businesses promote the idea you are not getting enough, or even too much. Understanding how much sleep you get, and of what quality, becomes a problem. You did not know it was a problem, but there are so many articles and adverts online convincing you that it is. Sleep becomes the reason you are underperforming at work or at home, a lack of sleep knocks days of your life and makes you more susceptible to mental and physical illness. Having convinced users that sleep is a problem, the digital world then sets about selling solutions.

Devices to monitor your sleep, connected to your phone,

that feeds sleep related data to a tech company. Products to improve the quality of sleep, quite possibly sold by or through Amazon or an advert that you see only because Google and others know you had been reading about sleep and are clearly worried about it. Furthermore, Valley leaders fund research into sleep and philanthropists find ways of helping the young or the poor to sleep better.

In a similar way other conceptual terms were adopted and either applied to user-focused applications or internally within Valley businesses and their employees. Brand-friendly, simple-to-grasp ideas like 'grit' - a sense of determination, perseverance and passion with which successful people, like those dynamic Valley leaders, are imbued and that less socio-economically advantaged people desperately need to discover within them. Or like 'growth mindset' - your skills and talents are not pre-determined by environment or experience but, like Human Capital, with the right investment anyone can achieve anything.

These concepts – not new, just newly defined for a modern, connected age - were quickly adopted not because of a concern within business for those who have been somehow deprived of the opportunity to develop these traits. Rather, it was because they are ideas that fit neatly into Silicon Valley's view of the world and of itself – they serve a form of Confirmation Bias. It became important to demonstrate grit, or employ those that did, and if you did not, that was your fault, but do not worry because technology can help you develop your hidden grit. An organisation's culture must encourage a growth mindset and ensure their people get enough sleep, and technology can help your organisation to achieve this. After all, these new, powerful companies were founded by those with grit, those with growth mindsets, resilience, creativity and personal leadership. These concepts, defined and shaped by thought leaders, became company assets. Products and services, inspired by these ideas, designed and sold to everyone from busy office workers to governments in order to develop these skills and traits. They also became philanthropic projects, which also aided research into new products.

This perspective has led to a commercial boom as well as a corruption amongst the tech companies. Users are less seen as consumers, but as assets of a company that they need to exploit. By constantly finding new ways to exploit them tech companies derive more data. People are the mines; data the gold. And like the Californian Gold Rush, most of the money and power remains with the hands of the wealthy and influential. With data acting as a new currency there came a new application for behavioural economics; the rational and irrational relationship with money could now be seen as applicable to data. Concepts like friction and the sunk cost fallacy were now relevant to data as well.

The desperate chase for data saw tech companies embrace elements of behavioural psychology. They employed more experts and looked to the latest research in the field and considered how their work could be used commercially. Emotion, cognitive psychology, biases were applied to draw users in, keep the data flowing, and nudge them.

At the core of Silicon Valley's new golden age was the sense of self. Each user should feel unique, special. They should be able to personalise. They should have a voice equal to anyone else. Technology could help confirm the adolescent solipsism that secretly stays with everyone, but that society usually suppresses. And in doing so users revealed more and more about themselves. Each item bought, video watched, comment made revealed something. Users were quick to believe in their own special, unique nature. They readily filled out surveys, believing their voices would be heard. They completed questionnaires in order to understand more about themselves. They shared the results with others who did the same.

The cognitive biases that are now a part of mainstream psychology were used to get specific, predictable responses from users. The Confirmation Bias (seeking out the information that confirms your already held opinions) and the primacy bias (the inclination to rely on the information most recently consumed) pushed users ever more into defined communities, the better to be targeted by advertisers.

These communities became part of tech companies' tool kit of segmenting users so that they can be targeted with messages that will appeal to them and only them. The more that specific groups are certain they are right, and have the weight of evidence to 'prove' it, the more they become separated from other groups. This isolation also means they become less rational - they are increasingly less exposed to other potentially relevant information, diminishing their good judgement. Paradoxically, however, despite being less rational in their judgements, they do become more predictable with tech companies able to build a more accurate map of where a user will go because the options are increasingly limited by their own online behaviour. They only go where others in their group tell them to go, avoiding anything that opposes or challenges them; sticking to what reinforces their world view, their experiences.

Whilst being able to manipulate users into giving up their data and also making that data fit certain useful models was all well and good, getting the data once was not enough. It had to be a long-term relationship between company and user. Change had to be monitored. Patterns of behaviour needed to be established over time. Extra information that might not have been found before could be acquired, but only if the users kept coming back, kept giving up data.

When computer games first came to market in the 1970s and 80s they were limited by computing power. Even the antecedents of home gaming, the arcade machines, could only do so much despite being big, expensive machines dedicated to doing just one thing. One way to keep players involved, to keep them spending in arcades, was to design the way the game worked. Yes, great graphics and an interesting idea helped, but if it was too hard or to simple, players would lose interest. The key was to design a game that rewarded repeated play. That each time you played you understood a bit more; progressed a little further. The next time you played, you would be ready for the combination of jump, run and duck that your character needed to make it past an obstacle, or the best place to position yourself in order to defeat the end of level boss. Each play gave you experience, you learned, and so the next time you progressed a little more. It was just challenging enough.

There is no financial reward to most video games. Yet players experienced a 'high' – a euphoria when completing a level or defeating an enemy. The high, to many regular players, was almost addictive. Video game reviews referred to the addictiveness of a game as a quality; a measure of whether it was worth your money. Being addictive, in consumer terms at least, became synonymous with being good. An addictive drama series; an addictive snack.

Personality traits such as self-control have long been a

mainstay of psychological research. However, neuroscience came to more fully explain this type of characteristic. Along with other aspects of who we are, it placed them in a context that defines them as universal to all humans. Studies into the physiological nature of the brain have revealed that certain parts govern certain psychic activity and are affected by the production of key chemicals in response to outside stimuli. In most cases a person cannot control the production of these chemicals by any means other than controlling the stimuli (that is to say they cannot physically control their reaction to it). The release of these chemicals is usually linked to fundamental, evolutionary needs like finding food or shelter or preparing to fight or flee in order to survive.

These chemicals, or neurotransmitters, relate to memory and general brain activity whilst also carrying out roles in the body such as engaging muscles and producing digestive enzymes. Within the catalogue of brain chemicals are serotonin and dopamine. Serotonin is generally referred to as governing mood and is released upon anticipation, heightening the sensitivity of other senses, making us feel alert when we know something is about to happen. Low levels of serotonin are linked to depression, a mood where activity and sensitivity become difficult. Dopamine has been dubbed the reward or pleasure chemical. It is also connected to emotional responses and heightens our reactions to being able to spot and seize potentially pleasurable experiences. Its release has been associated with everything from food to sex and is unfortunately also linked to activities including drug-use and gambling.

It is this latter area that has seen much work focused on understanding why addiction happens. Whilst some narcotics are physically addictive - that is, the body becomes reliable on them - mental addiction is far harder to tackle, in part because of the hard to control impact of dopamine. The evolutionary case for dopamine has linked certain behaviours with pleasure because they were needed for survival. Food, sex, physical security and social interaction had to feel good so humans would seek to repeat the experience. It has helped us survive hard times, it encourages us to give to charity, but it also makes us seek the approval of others – and not always for the right behaviours.

Dopamine is produced in the brain stem and then flows through the brain, affecting certain points on its way. As it makes its way it has an influence on parts of the brain governing learned behaviour, cravings, emotional responses, visual memory and imagination. Somewhat inconveniently the physical flow of dopamine means it tends to affect the impulsive or System One part of the brain first, making us more susceptible to, and less questioning of its rewards.

Humans, as they so often have, became their own worst enemy. Evolution had provided a physiological advantage, but humans started to find artificial ways of provoking these pleasure feelings. Whether it was smoking or gambling or consuming sugar or taking drugs, the production of dopamine on demand became something that could be sought out and, in time, commercialised. The combination of pleasure and learned behaviour saw addiction become a human phenomenon and one that was hard to manage not just because of the pleasure itself but because of the habits it also created. It is why addicts struggle with verbal or visual reminders of their addiction, or even certain experiences that they now link to the dopamine rush they once had.

Addiction, once seen as a moral failing, a weakness amongst some, is now understood to be a physiological characteristic, and in extreme cases, an illness. Neuroscientist Brian Anderson has also argued that addiction, or behaviour very similar to it, is not something someone is predisposed to but has the potential to affect anyone. We all possess levels of dopamine and the associated receptors in the brain; we all have the brain functions that can hardwire links between stimuli, behaviour and reward.

Further neuroscientific research is uncovering links to bullying and aggressive behaviour and dopamine release (amongst those doing the bullying, of course). Those that are in turn bullied or are or simply feel to be marginalised and disenfranchised, are more likely to seek out and become addicted to the dopamine hits of gambling, drugs, and even bullying itself.

Whether addicted or not, the use of drink, drugs, sex and food to stimulate dopamine has been around for centuries. A more recent dopamine producer has been observed. Being liked on Facebook or retweeted on Twitter or subscribed to on YouTube are all shown to stimulate dopamine production. Technology has, accidentally or not, found a way to affect us in a fundamental way.

Early video games were basic, and relied predominantly on their novelty for their attraction. But as processing power increased and games became more sophisticated and subtle in their design, so games makers started to incorporate a degree of psychology and neuroscience. Progress in games was made just difficult enough. Parts of a game, played enough times, could be mastered eventually. Players were taking on the computer, and when they won, the sense of achievement was no different to any competition. This is gamification and its use of reward and repetition has seen it applied extensively in the digital world.

Behavioural economics has shown how our rational views of risk and reward are skewed when it comes to money, something we value to a degree that it frequently makes us illogical and self-defeating. Part of this is down to the dopamine produced by financial rewards. However, our view of financial rewards change depending on how visible they are. If we purchase something others will see, we will often spend more or go further out of our way to acquire them.

Money is not the main currency of the digital world or the leading measure of status that it is the human world. The exchange of data is often a greater priority, and social interaction is the measure of an individual's worth.

Whilst the sense of progression and achievement in video games produced a pleasurable feeling, being able to point to one's position atop a high score table was even more rewarding. Social media uses likes and retweets and follows as a mechanism to keep score. The winner is the one with the biggest number; a number everyone can see. This drive for the biggest number sees people act in irrational ways, giving away data, using their phone when they should be driving or looking at a sunset or paying attention to other humans.

The psychologist BF Skinner developed an experiment involving a box; the 'operant conditioning chamber', later known as the Skinner Box. It used an animal, often a rat, a button, and the delivery of food, in order to understand learned behaviour and habits. In one test food was delivered at various intervals of button press. For a period of time, every press would deliver food. Later, it would be every 100th press of the button. The rats' behaviour would adapt over time, but remove the fixed interval and something happened. When the delivery of food became random, and the rat could not learn if food would come with the first, tenth or 200th button press, it just pressed the button more, appearing desperate, continuing to press in the hope food would come.

Similar behaviour to that displayed by the rats has been observed in another group: social media users. The constant notifications, each time usually of little consequence, and users, knowing full well that the messages are probably not important, check anyway. Studies show that whether a person checks the notification or not, their attention is affected by the knowledge that a notification awaits.

Dopamine anticipates a reward and tells the brain that it wants more of that thing. It motivates the rat to keep pressing the food button. It motivates the human to check if someone might like them or if they will learn something before others. It affects our mood, releasing pleasure when fulfilled and frustration and disappointment when it is not. It keeps us checking, posting and updating. It means humans react to likes, updates and follows in the same way as they might to food and sex.

The longed-for notification alert that has to be checked even though the odds are it is nothing (as in Skinner's experiment) and the like or follow that provides the rush are very modern addictions. Tech companies know when to alert and when not to. They know how to provoke a reaction. They know that the search for dopamine makes us behave in certain ways. Like advertisers, users aim for a reaction rather than honesty. That can lead to some dubious content designed to provoke rather than progress.

Dopamine also affects views of risk, as seen by those stimulating dopamine production by undertaking high-risk activities from gambling to extreme sports. It exaggerates some of the biases and behaviours that behavioural economics has observed and described. It sees people do things they would not normally do, taking greater risks in the digital world than they would in the human world.

Skinner believed that the principles of behavioural control could, indeed should be applied by a ruling intellectual elite to create a more law-abiding, productive society. He also believed that positive reinforcement had the power to improve the world. The idea of rewarding certain behaviours has long been accepted as a way of modifying behaviours, from training a dog to educating a child.

Addiction has become a part of the Silicon Valley business model. Euphemistic words like 'sticky' are used, but it amounts to the same thing. The more time a user spends there, the more data, and in some cases money, can be harvested from them. Every generation has warned about the corrupting nature of spending too much time consuming everything from television to pulp novels. The difference in the digital age is that the online world follows us around all day, feeding our lowest instincts in new, ever-changing, ever more insidious ways.

User profiles and data, complex, connected platforms, ideas that create and feed addictions – all of these new, unprecedented, intangible assets make ownership hard to define. As a result, lots of highly skilled people are employed in order to assert and maintain ownership over these assets. In doing so, it brings tech companies into closer and closer contact with law-makers and regulators. Cities draw in more and more people under the gravitational pull of power and money. There develops a spiral of ever-growing importance and complexity. And with greater complexity comes a greater pressure on government to manage it.

Our irrational nature, our search for the approval of others, our malleable behaviours, our addictions. Humans once again need saving from themselves. But in the non-geographic world of the digital, what power do conventional institutions have? Who can they turn to in order to manage these complex, irrational populations?

PART 4

PREDICTING THE FUTURE

With economic prediction seeming, at least the mind of the general public, irreparably damaged, what would fill the space? Perhaps the best predictors are those shaping the future themselves, and with access to unparalleled knowledge of, and influence over, almost everyone. Technology and technologists had become dominant in commerce and would be increasingly powerful in government and society.

A technologically driven future holds many concerns, amongst them the effects of reducing people increasingly to generic data. What inspired technology, technically and culturally? What insight could be gathered from that? How would artificial intelligence, key to the future of mass data processing, alter prediction and how it is seen? Could technology make humans more predictable? And if so, at what cost?

19

VISIONS OF THE FUTURE

Humanity's need to plan and prepare, whether to protect themselves or to gain advantage, has led to a desperate search for reliable predictions. Weather, natural disasters, economic fluctuations, the outcomes of conflicts or gambles - all have been analysed by means that are superstitious, religious, scientific, and quasi-scientific. These predictions have been considered, at the time, vital; of national or even international importance. Money and resources were poured into making them as reliable as possible. Luck and skill, manipulation and intelligence played a role in elevating those that predicted into well-rewarded, fated individuals.

This history of prediction is littered with misconceptions and mistakes. Sometimes very costly mistakes. There is a group, however, whose predictions are considered generally less realistic but are no less influential for that. This group have, indirectly, shaped technology and society often far more than any economist or political or religious figure. Their predictions have had little of significance resting on them, so accuracy was of little importance, but those predictions have been retrospectively interpreted as being remarkably accurate, especially considering the timeframes they worked to – often projecting decades, even centuries ahead.

Visions of the future have played a role in creative arts for centuries. Artists have taken technology, the possibility of other worlds, and humanity's ongoing search for god-like abilities as inspiration. Such notions have influenced music (particularly those working in experimental and avant-garde forms of classical composition, as well as those using new technologies to create more popular forms) and visual arts (most notably in the Futurist movement). It is, however, in literature, and later performance media (film, television, and to a lesser extent theatre) where visions of the future have found a natural home.

The origins of science fiction as a literary genre are debated, with cases being made for it dating back to 2000BCE with the *Epic of Gilgamesh* (the king that would be immortal). From around 400BCE there are examples in Greek, Indian, Arabic and Japanese literature of manned flight, travel through space and time, other worlds and robots (although that term would not be used to refer to human-like machines until 1920).

The robot, along with space, and travel to other worlds and eras, have been mainstays of science fiction since the genre emerged. Both were really only believable with advances in science and technology. As a result, such literature moved away from fantasy and the metaphysical and towards a grounding in reality; to extrapolating from the latest scientific developments and asking what will happen when those developments become commonplace and even obsolete.

The genre's reliance on technological advances, both real

and fantastic, meant it was only in the 19th century that science fiction began to find its voice. Taking direct influence from real scientific endeavour it merged human stories with engaging, novel ideas and the use of often-cautionary metaphor. Mary Shelly's *Frankenstein* is generally accepted as the original pioneer of science fiction despite the expression not being used for another hundred years. The years following its publication saw a number of writers speculate on the future or where the new scientific age might lead society. The two giants of 19th and early 20th century science fiction, and arguably those that came closest to formalising the genre, were HG Wells and Jules Verne.

Wells' works includes *The Time Machine* and *The Island of Doctor Moreau*. These titles enabled Wells to ask questions about contemporary social order, conflict, and humanity's dominance over other species and where these questions might lead. In his non-fiction *Anticipations of the Reaction of Mechanical and Scientific Progress Upon Human Life and Thought: An Experiment in Prophecy*, the author, to some degree or another, predicted a world at the end of the 20th century built around cities filled with trains and cars and commuters. He hinted towards the sexual revolution of the 1960s, the EU and an America-dominated world. But he also saw the end of capitalism and democracy (replaced by a form of technocratic, expert rule), the reduction of world languages to just three or four, and, common with many writers and thinkers on all sides of the political spectrum at the time, suggested some form of eugenics would alter the human race for the better.

Verne's work, almost 50 years before Wells', was more representative of a stylistic shift from fantasy to science fiction as befitted his age. He laid many foundations for the genre allowing others after him to develop new ideas. *Twenty Thousand Leagues Under the Sea* focuses on the scientific brilliance of Captain Nemo and his quest for both knowledge and revenge on the society that excluded him. *Journey to the Centre of the Earth* also focuses on the questionable, erratic brilliance of one man whose personality turns science into something both inspiring and dangerous. Verne is more concerned with the human than the technological (except, perhaps, for his electrically-powered submarine), but away from his fiction he did predict a form of global news that was broadcast for mass-consumption.

Both Wells and Verne imagined a time when travel to the moon was viable, even if they could not quite accurately forecast the means, context and implications. Verne even included solar-powered space transport and space travellers landing in the sea (foreshadowing the splashdown method used by the US and USSR throughout the 1960s and 70s).

What Wells and Verne hinted at - the social and political implications of technology (be it an over reliance on it, or an ignorance of it) - others would take as the core purpose of their writing. Science fiction would increasingly use fantastic ideas of a future world to reflect on the nature of contemporary society and politics. For George Orwell and Aldous Huxley a possible future offered profound warnings about the present. They created two of the early 20th century's best-known literary visions of the future in *Nineteen Eighty-Four* and *Brave New World*.

Nineteen Eighty-Four has become as famous as a cultural phenomenon as much as a novel, partly because of its dazzling originality, but also because of Orwell's use of language to describe concepts that at the time were complex and far-fetched but that have since entered common use. *Nineteen*

Eighty-Four, written in a world where fascism had given way to a form of Communism that to many looked very similar, is noted for two significant observations. One being the use of disinformation; the power that controlling the public's access to news and information can bestow. The other being the idea of subservience through surveillance; equality through fear. Both of these were evident in Nazi Germany but were particularly pernicious in the Stalin-era USSR.

Although contemporary to the late 1940s, Orwell's vision has repeatedly been cited as prescient in the West in every decade since. It seems that every generation thinks their society is more like the world of *Nineteen Eighty-Four* than the last. It has never been a flattering comparison. Governments are accused of changing history, of misleading, of controlling, and worst of all, of monitoring. A screen in the home both feeds 'news' and observes the viewer. There is distrust between neighbours. There is a remote, shadowy ruling elite. Facts are no longer facts unless certain people say they are (but those people are never known). Simplistic, base entertainment is used to distract the public. Unthinking nationalism, simplistic and often violent solutions to social problems, euphemistic political language (indeed, the general abuse of language to confuse, undermine and obfuscate – something Orwell was particularly alert to) are all part of Winston Smith's Oceania. For Orwell these were desperate warnings. For contemporary readers of all political persuasions they are a vision of what is just around the corner.

Orwell was from an upper-middle-class family of landowners and mid-level colonial diplomats; a background he largely rejected both personally and politically. His fellow Old Etonian (and one-time teacher) Aldous Huxley was from a family of intellectuals and scientists, and he embraced this, along with a contemporary taste for satire and iconoclasm. Huxley's *Brave New World* is perhaps less popular than its more overtly political younger cousin, but it is no less important. Indeed, many argue it is more relevant and prescient than Orwell's tonally bleaker vision.

Huxley's world is sleek and industrialised, consumer-centric and deliberately American – for example, the calendar counts from not the supposed birth of Christ but from the year in which Henry Ford produced the first Model T. Humans are created to order and according to society's needs - the right proportion of dull workers to superior leaders to middling middle-managers. Here too, language is used to mendacious political ends, and there is a shadowy ruling elite (the ten World Commanders). However, whilst *Nineteen Eighty-Four* gives a sense of oppression and gloom, *Brave New World* actually implies a content existence. Happiness through conformity – doped and indifferent to the inequalities. A society subdued through luxury, entertainment and the consumption of an ecstasy-like drug, soma. A society comfortable knowing that the best people for the job rule the world because they were created as such.

At a more mundane level Huxley predicts genetic engineering, forms of entertainment that operate beyond sight and sound, personal flying transport, and a (largely unremarked upon) method of preserving one's youth. Fundamentally he sees a world where the pursuit of progress and reliance on technology leads inevitably to being controlled by, and subservient to that technology.

There are a number of parallels between the two books. Both have been regularly cited as maps showing the roads contemporary society has already started down. From CCTV and the internet to increasing automation and concepts of morality, they are deemed to be warnings that if we do not stop this now, if we do not regulate or rise up in opposition, a horrible future awaits us.

It is, however, society's relationship with itself and its leaders that hold the direst warnings. Both *Nineteen Eighty-Four* and *Brave New World* predict a stratified society where your role in the social order defines you and cannot be changed. To question this is to challenge society as whole. Society itself becomes a tool of the oppressive regime. Around two decades before Stanley Milgram explored the human unwillingness to stand out from the crowd, Orwell and Huxley were warning of the dangers posed by any society that views free thought as suspicious, even treasonous. In both novels the individual was worthless yet feared by the rulers. They were powerless yet had the power to disrupt the well-ordered world the rulers valued.

Where Orwell saw a state that would break the freethinking individual so that they conformed, Huxley took a different view. Those seeking to question the prescribed behaviours were sent to live outside of the attractive, comfortable, modern cities so that they could not disrupt the modern world. Their punishment was discomfort and solitude. Individualism could only exist in exile. The price of comfort and community was conformity.

Huxley later stated the world was heading more quickly towards his *Brave New World* than he expected. Twenty-five years after he wrote the book he felt psychological conditioning, drugs and overpopulation were paving the way to his vision of a luxurious, ordered dystopia before the end of the 20th century.

Brave New World was a reaction to the outcome of the

First World War, and *Nineteen Eighty-Four*, a reaction to the Second as it moved into the Cold War. In the years after the dropping of the atomic bombs on Hiroshima and Nagasaki, science fiction became both respectable (it had fallen prey to being written off as trashy or sensationalist) and urgent. It remained a political metaphor, as writers would grasp the Cold War analogies of alien races or robots enslaving humanity to underline the importance of anti-Communist tactics. Other writers recounted the dangers of placing scientific and technological advances solely in the hands of the military and the politicians, especially in a world where the destruction of humanity was not just possible but apparently quite likely.

As well as warnings, writers were also looked to for inspiration of what might be to come. Creative thinking, rather than logic and strategy, provided a vision – something to strive towards or prepare the world for in the longer-term.

As science and technology both advanced and became less the concern of academics and more an everyday reality, science fiction writing evolved. It continued to ask humanity difficult questions whilst also looking ever more to the science behind the future. The 1950s is seen as the golden age of science fiction, and of that era three writers – Isaac Asimov, Arthur C Clarke, and Robert Heinlein - known as the 'big three' - dominated. It is Clarke and Asimov who are now widely seen as the most important, enduring, astute and influential, well beyond the worlds of literature and film in which they worked.

Clarke and Asimov studied science and technology; they immersed themselves in these subjects as well as areas of social science, in order to create visions that were both compelling, and convincing. Visions that were less about fantasy and more about what might really be possible if humanity took a particular route.

SCIENCE FICTION/SCIENCE FACT

"No one can see into the future. What I try to do is outline possible 'futures' - although totally expected inventions or events can render predictions absurd after only a few years. The classic example is the statement, made in the late 1940s, by the then chairman of IBM that the world market for computers was five. I have more than that in my own office." - **Arthur C Clarke**

Clarke wrote that quote in 2001 - the year that provided the title to his most famous novel. In fact the novel was actually a film first, *2001: A Space Odyssey*. Co-written with Stanley Kubrick, *2001* is an ambitious, perhaps too ambitious, attempt to realise a history of humanity from the early hominids to a future of space travel. It draws parallels between man's dominance of the world through technology, and its dangers. Whether that technology is a bone used as a weapon, a nuclear arsenal, or a computer used to sustain life where no life should survive (in this case, deep space). As well as its grand scope, originality and visual drama, the film is notable for seeing a 21st century featuring the innovations of videocalls (another regular feature of science fiction – writers have long assumed we would want to see each other as well as hear each other) and artificial intelligence.

Although he was the author of over a dozen science fiction novels, Clarke became more than a writer of fantastic stories; his ideas established him as a sort of public predictor. Not of the rise and fall of nations, of economic markets, sport or weather, but of technology and its role in everyday life. He outlined the technological marvels in store, 20, 30, 50 years hence. He was a regular on TV screens and was even well known enough to have a number of series featuring his name, including *Arthur C Clarke's Mysterious World*. He is often cited as one of the 20th century's best predictors. He once said that the less believable his ideas were, the more likely they were to turn out to be, at least partially, true.

The list of developments Clarke foresaw are long, varied, and in some cases require quite liberal interpretation. He foresaw online banking and shopping (of a type), and the internet in general (actually a library of information that can be accessed instantly ranging from news to flight details). He described something akin to a mobile phone, although more interestingly he suggested such a device would radically change society. Longer-term he predicted public space flight, a universal currency based on energy (not dissimilar to cryptocurrencies whose value is influenced by the processing power required to 'mine' them), dinosaur cloning, the end of work, and human-level AI.

In 1945 Clarke also speculated on communications satellites (he would give his name to the concept of satellite geostationary orbit) over a decade before the first ones were successfully launched. He suggested this would result in instant communication between anyone in the world, even without knowing where they are, a prediction that only really started to become true in the last decade or so. With this death of geography, one could live and do business anywhere without disadvantage. It would also mean that one day a surgeon in Edinburgh could operate on a patient in New Zealand. By 2001, this was a reality (French surgeons in New York performed a gallbladder removal on a patient in Strasbourg) if not commonplace.

This last example typifies many acclaimed predictions from Clarke and others – there are uncanny elements of how events really have panned out, but also important shortcomings. Clarke had the insight (and luck) to see, decades before the technology was even near realisation, that a human would be able to control a robot thousands of miles away with pinpoint precision, as well as a beneficial, life-changing application for it. At the same time, he did not concern himself with the hows and whys, the actual steps required to get there. The fundamental changes in microchips, motors, the move to digital technology, and human-computer interfaces. Nor, at least in this case, did he pursue the consequences.

The consequences of only needing a few brilliant human surgeons around the world, but many robot surgeons, has huge implications for everything from the cost of training to patient outcomes. One advantage of having lots of surgeons is that if one gets it wrong, only one patient will suffer. Who gets sued if the robot gets it slightly wrong – was the fault of the human, the machine, or the interface?

Some of Clarke's predictions feel still somewhat wide of the mark. He proposed that the dominant intelligence would come to be electronic, starting a non-organic form of evolution. Although he also suggested that rather than humans employing robots to carry out dull and menial tasks apes would be the new servant class. He suggested all matter would be manipulated at an atomic level and machines (in some ways akin to 3D printers) that could replicate anything. He expected domestic nuclear generators and the scrapping of nuclear weapons (amid mass public revulsion following a third aggressive deployment).

Perhaps his greatest advantage in the prediction stakes

was that Clarke inspired as much as he predicted. By painting engaging pictures of incredible technological achievements, many based around real, cutting edge developments, many scientists and innovators actually strove to make his visions a reality. By issuing predictions that would influence the chances of them coming true, whether he lived to see it happen or not, he greatly increased his odds of success. He was broad enough to be open to interpretation, but where he was more specific, it was because the work in that direction was already underway.

That is not to suggest some cynical method or motive on Clarke's behalf, but more a commentary on the public reaction. The desperation of people to know what is to come. Clarke sought to inform, to inspire a passion for science and technology, as well as to warn. However, like all artists his goal was ultimately to entertain. He was not paid by governments or businesses (although they may have studied him) to promote or endorse their agenda. He could be free-thinking so long as he was engaging. He did not have the pressure of investors' millions riding on his far-fetched ideas; no government was enacting a policy affecting millions of people on the basis of his advice.

As Clarke noted, these big picture predictions are founded in science and inventions that were at the time contemporaneous. Working in the age of the Space Race meant he understandably believed space held the future, whether it was new colonies, new mineral resources, or simply the motivation for new technologies.

In 2001, when Clarke commented on IBM's lack of foresight in predicting just five computers in the world, he went on to say that having incorrectly predicted a Mars landing by 1994, it was now unlikely to be before 2010. At the time of writing (2018) the big recent shift in space exploration is the move from state (Nasa, ESA, Roscosmos) to private funding. One of the leaders in this field, Elon Musk and his SpaceX are aiming to land humans on Mars around 2024. Musk himself claims to be inspired in his Martian goals by works by the other titan of 20th century science fiction, Isaac Asimov and his *Foundation* Series.

In Asimov's *Foundation* stories, the ultimate hero (despite being dead for much of the duration of the tales) is Hari Seldon. Contrary to a convention owing its origins to classical literature of the heroic archetype, Seldon saves humanity through science and mathematics, rather than bravery, strength or strategy. This has made him a heroic and inspirational character to many serious-minded science fiction fans who have gone on to try, in some way, to emulate him.

Seldon develops a truly scientific form of social science with hard, infallible maths at its core. It is a way to both predict and shape society, and in shaping it make it predictable. This discipline, 'psychohistory', predicts the coming of a 30,000-year age of social disorder that, whilst unavoidable, can be reduced to a mere 1,000 years. The solution lies in the development of a special group, The Foundation of the title. The Foundation are a community of uniquely talented people – thinkers, artists, engineers - charged with guarding and expanding human knowledge. To protect them they are placed at either side of the (populated) galaxy. (Elon Musk, incidentally, agrees that humanity's best option for preservation lies in colonising other planets). Throughout the long and dense series of books and short stories, Asimov outlines the evolution of the socio-economic state and the positive effects of psychohistory. Battles are not wars, but psychological experiences; technology is synonymous with religion; economic and intellectual power trounces any other form of power (there is a general distain for both inherited power and martial power throughout the stories).

The *Foundation* series represents a complex, ambitious attempt to use science fiction to reflect on humanity's progress. Political, technological and social in scope, Asimov avoids condemning or condoning actions or individuals in his world, merely stating what might be. There are few outright heroes or villains and it is groundbreaking in using a whole new branch of science as a central plot device. It might inspire some to colonise planets or seek out new predictive methods, but for the more casual consumer, it is too challenging to really influence their ideas of the future.

SEEN TO BE BELIEVED

From its inception film felt futuristic. The very idea of seeing a moving image, of seeing what was not there, was both terrifying and exciting. Technology was changing how people saw the world and made them question what was real. Just as visual artists could create still images of the impossible in paint and photography, film presented the possibility of making the impossible seem alive, animated. Science fiction was an obvious early source of stories, enabling a vision of a future to be seen as well as imagined.

Within ten years of the invention of cinema, film pioneer Georges Méliés created the first science fiction film, as well as applying the first use of special effects. His 1902 *Le Voyage dans la Lune* (A Trip To The Moon) was inspired by Wells' and

Verne's imaginings of human lunar exploration.

As film technology improved, feature-length, sophisticated work appeared; audiences were more comfortable with film as a way of telling stories rather than it just being a spectacular experience in and of itself. *Metropolis*, the 1927 Fritz Lang and Thea von Harbou film is often cited as the most visionary of early science fiction films. Serious, politically-charged, stylistically brilliant, it portrayed a future of mass industrialisation and stark social division, with a powerful ruling elite and exploited workers, as well as a robot that can pass as a human.

From thereon in directors attempted to envisage a future that was both entertaining, inspiring and thought-provoking. Whether original or based on literature, film would enable artists to realise a vision in a way no other media could. The nature of film, how it could be manipulated, how it could make models, painted images, props and other effects look real, made it the ideal playground for science fiction and visions of the future.

Film came of age either side of the Second World War, so it is no surprise that science fiction's application as a political metaphor should seep into film. The 1940s and 50s saw science fiction frequently used as a Cold War metaphor, often seeing alien races seeking to dominate wholesome Americans. As such the settings were often contemporary, rather than futuristic, making the threat of outsiders enslaving Americans all the more real. There was also a strain of lower budget, horror-infused films designed to scare and thrill, and adventures aimed at young and family audiences.

By the mid-60s both film and science fiction were more respectable, and films such as *Fahrenheit 451* and *2001: A*

Space Odyssey brought a strain of intellectualism alongside the drama. In the years following the Cuban Missile Crisis the threat of nuclear Armageddon felt more real than ever. Science fiction looked to what would happen in the wake of global destruction as well as the hazard that science and technology now represented. The 1960s saw starker, dystopian visions like Jean-Luc Goddard's *Alphaville* with its backdrop of a brutal and brutalist, technological society. The film maintained the theme of an individual trying to fight the inhuman oppressor, this time a sentient computer, rather than a human or alien dictator.

Cold War paranoia remained all-pervasive in the 1970s, and science fiction was becoming ever more serious in tone and theme. With science fiction film-making invariably requiring bigger budgets it was dominated by Hollywood, but in the Soviet Union director Andrei Tarkovsky brought subtlety and intellectualism to the genre. Based on a Polish novel, Tarkovsky's *Solaris* was a meditation on psychology, isolation, perception and communication with others told in a style and pace that was unexpected for the genre.

In the US, science fiction started to reflect a strain of neurosis born of political and social strife as well as the birth of a surveillance state. It found a ready domestic parallel in Nixon's White House, renowned for its paranoia and its surveillance both internal and external. Bleak, apocalyptic visions with an unsettling edge and, by Hollywood standards, uncertain or outright downbeat conclusions became the order of the day. For example *The Omega Man* saw humanity trying to survive in the aftermath of a biological war and the resultant battle between those that felt they had been doomed by technology and the minority who believed it held the key to saving them.

Soylent Green saw a vast mega-corporation conspiring against society, and *Westworld* saw robots built for human exploitation and entertainment (without moral consequence) turn on their human creators.

The 1970s also brought a new era in film via science fiction in the shape of *Star Wars*, which would change what science fiction was in the minds of the public for years to come. *Star Wars* was arguably the first 'event picture', something spectacular, big budget, and multi-media that operated beyond the screen and the box office. Although culturally important, it was no vision of the future. What it did was allow its director George Lucas to establish a new age of special effects which would allow other visionaries to create ever more convincing worlds and ideas.

Off-screen the billion-dollar entertainment industry in Hollywood drove technological innovation in neighbouring Silicon Valley, which in turn facilitated and inspired ever more spectacular ideas. In 1973 two computer science students, Ed Catmull and Fred Parke created the first commercial CGI (computer generated images) for the film *Westworld*. In 1975 Lucas founded Industrial Light & Magic, a pioneering visual effects company primarily designed to work on his *Star Wars* project. In 1979 Lucas recruited Catmull to head up ILM's computer graphics division and six years later Catmull started Pixar, with initial investment from Apple's Steve Jobs. By this stage computers were starting to be relied upon ever more for complex, audience-stunning creative visions.

Largely under the influence of *Star Wars*, many of the science fiction films of the 1980s were big-budget spectacles focused more on entertainment than on examinations of possible futures (though in some cases no worse for that).

Although not a huge box office hit at the time, especially for its budget, *Bladerunner* crossed some of the divide between mass-market entertainment and intellectual rigour. Inspired visually by *Metropolis* as well as film noir of the 1940s, *Bladerunner*, based on Philip K Dick's novel *Do Androids Dream of Electric Sheep?*, merged Hollywood style with European sensibilities. Although ostensibly a thriller it explores moral philosophy and religious themes in a world where humans have become gods through their creation of human-like androids (replicants). Created solely to do dull, dangerous and dirty work, they are nonetheless superior to humans and have started to question their place in the world. This promethean theme is something director Ridley Scott would return to in his other science fiction films. In particular, *Bladerunner* questions memory, consciousness and the soul and what it means to be sentient which, in 1982 was well ahead of its time. Once again, viewers see mega-corporations, consumerism and superficial entertainments governing the lives of a large underclass.

The post-Cold War 1990s saw nuclear conflict invariably replaced by environmental catastrophe (a reflection of real-life concerns) as the prime threat to human existence and catalyst to change to the rules of morality and survival. The decade also saw the internet become a source of fear and of hope. Another Philip K Dick story, *Minority Report*, was adapted by Steven Spielberg in 2002. It looked fifty years hence to a time when law enforcement relies not on conventional crime detection and prevention, but on a system that 'sees' that a murder will happen and imprisons culprits ahead of them perpetrating the offence. It envisaged a future in which, not unlike Asimov, prediction underpins the safe operation of society (albeit predictions that rely on the foresight of mutant creatures rather than a newly discovered science). Arthur C Clarke suggested that by 2010 a 'Big Brother' system of surveillance would have essentially eliminated crime. *Minority Report*, in all senses a big Hollywood film, reframed many established science fiction themes for the post-Data Revolution age, not least questions of privacy and surveillance and of freewill and governance in a technologically dependent society.

As the 21st century progressed, so did real-world advances in artificial intelligence and automation. It saw films like *Bladerunner* become an increasing influence over new science fiction stories. Just as it had asked questions about the nature of the soul, consciousness and sentience, so a new generation of writers and directors used the ubiquity of technology in modern life to bring such ideas into a more recognisable near-future. It made questions about what it is to be human and how new forms of 'life' should be treated all the more urgent. Films like *Moon*, *Ex Machina*, and *Her* took seriously the trajectory of contemporary technology and its potential to replicate, even to replace humans. They asked audiences to think about how much of their lives they were giving over to technology, how their data might be used to predict what they want on an emotional level, and the relationships they had that would once have been with a human interlocutor. Audiences were invited to look at humans and technology and reflect on what the difference was.

20

WHO GOVERNS IN THE DIGITAL WORLD

Canadian thinker Marshall McLuhan was one of those people who saw the interconnection of the world as leading to a collective identity; a global village. In 1964 he coined the phrase 'the medium is the message' as a call to people to think more about the medium that carried the information than the information itself. A few years later, alongside designer Quentin Fiore, McLuhan published *The Medium is the Massage*, which further argued that, when viewed historically, it will be the technology (the medium) - whether that is a book, an item of clothing, or a new technology – that actually tells the story, not the content it carries. His use of the word 'massage' in the title has been taken as referring to the soothing, unchallenging nature of many media and their messages that serves as a distraction from the problems they in fact create.

In the digital world, humans have undoubtedly become

addicted to the medium, if not the message. It is undoubtedly also a medium that is reassuring and unchallenging. But does that mean the message is irrelevant? That in decades it will be the evolution of technology that has most affected humanity? Did cave paintings matter as much as the cognitive and physical ability to make them? Did the Qur'an matter as much as the ability to read, write and print a book?

In the long term, humanity may be freed from the subtleties of messages, but in the short-term we are governed by the digital conversations taking place at all levels, in all languages, from all perspectives. Once messages were one-way. They came in the form or laws (religious or secular) or pronouncements, holy books, speeches or censuses. Conversation only took place on an individual level. The voices heard were only of the privileged, and sometimes even the qualified. Geography and language largely dictated who heard what message. In a digital world, those rules no longer apply. Debate can be shaped by those not only unqualified to shape it, but with malign intent.

In Plato's *Republic*, the philosopher reflects on justice, human nature and virtue (which had a much wider, more nuanced definition than it does today). Within the ten volumes that make up the title is the story of a shepherd, Gyges, in the service of the king of Lydia. An earthquake uncovers a cave, and within Gyges finds a corpse which wears a gold ring. The shepherd discovers that the ring has the power to make him invisible (Tolkien readers may find this familiar). He uses this power to seduce the queen of Lydia, murder the king, and rule in his stead.

In *Republic*, this parable serves to demonstrate that morality is a social concept. Remove the social oversight, the expectations and judgments of others, by, say, being invisible, and a person will eventually no longer act morally. This could be seen as cynical; as taking a low view of humanity, but it is hard to deny.

Ancient Greece is often cited as a shining beacon of progressive thought; a guiding light for humanity during a collectively dark period. It 'invented' democracy by being the first Western state to incorporate the citizenry into decision-making. Yet, in 427BCE, a hundred years before Plato, ordinary Athenians voted to slaughter and enslave thousands in the town of Mytilene as a punishment for taking Sparta's side against Athens in the war. The next day, appalled at what they had done, they changed their minds – in fact, they debated intensely, eventually deciding that mercy would have a better long-term outcome by avoiding future rebellions. Although the ship carrying the order to kill had literally sailed, a second message arrived in time to prevent the massacre.

People are no more reliable or reasonable two and a half millennia later. Yet we are arguably more susceptible to seeing inconsistency in our leaders as a sign of weakness rather than wisdom. This ancient era of political debate brought about the term parrhesia. The word has acquired a few, albeit similar definitions over the ages. Some suggest it means to speak reason and truth, particularly to the powerful. It could mean that freedom of speech demands equality of speech - that all voices should be heard without favour. Or that it suggests asking forgiveness for the freedom of one's speech.

Not being open to the views of others, to debate reasonably on the basis of evidence, usually means a drift towards becoming insular, eroding empathy, and dismissing views that are difficult or different. It means that decisions are not robust, and are made poorly, and are likely to be regretted. This is what happens when a community insulates itself from challenges, new ideas, and outsiders.

In the US, a survey of driving habits (performed by observation rather than public polling) found that the most inconsiderate drivers were those in newer, black Mercedes. It is not worth drawing too many conclusions from this; black Mercedes are commonly used as up-market town cars with drivers more likely to be on the road for long periods and under more pressure. A less scientific survey in the UK suggested it was BMW and Range Rover drivers that were the worst for cutting people up, failing to indicate or pushing into queuing traffic. Particular car manufacturers and their target markets aside, many drivers report feeling more aggressive when behind the wheel. Although somewhat anecdotal, what this does suggest is that, isolated in a large, expensive car, protected by metal and safety features, closed from the outside world, feeling empowered by all of those controls and displays, the driver suddenly becomes a very different person. A more arrogant, less considerate person that is quicker to anger, and likely to take greater risks. Perhaps one that is ready to reveal their darker side; a side usually kept under control by social expectations or fear of judgement.

The social media user, holding more computing power in their hand than, famously, put two men on the moon in 1969, feels empowered. They can access any piece of information, they can create, they can automate, they can communicate with anyone. Yet they are also, like Gyges, invisible; powerful yet unaccountable. They can speak and speak loudly, but neither the audience nor the subject of their discourse will see them. They will not understand their feelings, their experiences.

Social media effectively dehumanizes both speaker, subject and listener; reducing them to a set of one-word traits, likes and dislikes.

This has become the dangerous currency of social media. People reduced to a few key words, devoid of subtly and understanding or anything that might inspire empathy. They find their community through these words and then share similar experiences, similar ideas. Their Confirmation Biases convince them they are right, and all others are wrong. When challenged, when they are forced to hear something they find disagreeable, they react with rage, in a way they never would face to face. They threaten and lie and insult.

Meanwhile there is conventional, traditional, slow-moving government. People in large buildings making rules intended to serve as many people as possible as fairly as possible. Institutions that have grown up over decades, even centuries, with the responsibility to manage groups confined and defined by geography. These people, these flawed, hard-working humans, have a new, constantly shifting and confusing challenge.

Government maintains an interest in competition; in not wholly embracing the concept of a global village. Their power lies in national borders; in differences of culture and economics between the manufactured concepts of nationhood. Government clings on to trade and treaty. Nationalism has a great deal of appeal in modern politics as a way of reinforcing differences and thereby encouraging competition. Politicians will seek to convince voters that their nation and its population are different, special, and unique. That only that politician or party can protect the nation and help it prosper. That only they know how to get ahead of other competing nations that are somehow both inferior and threatening.

People are increasingly used to being told they are unique and special. Social media thrives on it. Silicon Valley has it embedded into its being. American Cold War philosophy insisted it was true and, apparently, triumphed on it. So it is perhaps to be expected that the idea should convince so many people.

Except that many younger, more digitally-absorbed people reject these ideas of national politics. It is dated, old-fashioned, and worst of all, slow. A new generation does not just reject partisan domestic rivalries but the very idea of left and right. Technology has made them irrelevant. They have more in common with like-minded, similarly educated people thousands of miles away on another continent than they do with their fellow voter in their hometown. Living so much of their life digitally, they also feel more connected to their foreign cousin than to their compatriot. Ideologically they believe that entrepreneurialism and individualism (right) are not incompatible with sharing, mutual support and equality (left).

Communities, the groups made by tech companies to segment, manage, analyse and ultimately commodify people, appear to be the new politics. The members believe that their community holds the solutions to whatever problem faces society. Their solutions, however, tend only to apply to their community. They do not take into consideration the other communities that geographically if not digitally, they live alongside. Government has to work geographically; it is bounded by geography, exists because of it, and cannot regulate across boundaries for a chosen stratum of global society. Government needs to work for all opinions, backgrounds, ambitions, genders, races and experiences that fall within its geographical jurisdiction.

In a digital world, what is the role of government? Government must hold on to its position of running the world, regardless of how much evidence there is to the contrary. Whilst ideology may have had its authority eroded over the last 30 years, government still maintains a managerial role. A role that, over time, it has gradually improved at, and even become surprisingly adept at.

Whilst critics will always find fault, frequently on ideological grounds, government has proved gradually better at ever more challenging managerial tasks. In organising funds and resources and people it has demonstrated, historically, an improving trajectory. Over the decades more of the population has gained access to education, employment, healthcare, transport, housing and the essentials of life that government are, ultimately, responsible for. Whilst government policy and ideas may be flawed or disputed, their implementation has often been efficient.

It is hard to say if politics has seen an influx of a management class, or whether government's increasing role in management has made managers out of politicians. Either way, if management is to be the primary role of government (whether it wants it to be or not), it needs to account for vast, complex structures, competing economic priorities, widespread social changes and specific individual, personal needs. The key to this type of large-scale management is planning. The key to better planning, especially in slow-moving government, is better prediction.

THE BUSINESS OF GOVERNMENT

In Chile, in 1970, an openly Marxist president and a socialist government were democratically elected. The rule of Salvador Allende would end within three years. The country would fall under the fascist rule of the US-backed General Augusto Pinochet after a CIA-orchestrated coup, another turn in the road of the Cold War. During that short experiment in socialism, Allende implemented a remarkable, groundbreaking project: Cybersyn.

The project, drawing on expertise from around the world, was based on the latest management thinking, as well as the latest technology, and attempted to create a system that emulated the action of the human mind. Cybersyn's goal was to maintain the national economy and social order. It connected businesses and government to economic modelling software and a human decision-making body. It took in productivity data, weather, worker absenteeism and more, all in an attempt to forecast shortages, crises and movements. It was fed, produced predictions, and government acted accordingly. As short-lived as it was, had it survived, it may have stumbled not over a matter of ideology but over the rational nature of humans whether they live under capitalism or socialism.

Anyone charged with governing a whole population has many complex challenges, but economics has come to dominate. Effective economic management relies on seeing changes, predicting their effects and preparing accordingly. Whether it is an emergency aid package, a press release, an allocation of staff or an appeal to business. Data collected from a vast array of national and international sources at every level of public and private activity is needed. Data through which to comb and find elaborate, complex patterns. From there come predictive models in which to feed current data. Predictions are made, and government can decide what policy to enact, on whom, when, and for how long in order to have a desired effect.

Economics changed under the influence of technology - specifically with increases in computing power. In the years after the Second World War economics evolved hand-in-hand with technology. The move towards a reliance on maths and mathematical models demanded greater computing power and more data required more processing capacity. The conclusion in economics was that more data would produce more accurate models. Moore's Law fed technological advances with which economics could not really keep up. Economics surrendered itself to the precision of technology and the certainty of science.

Economics is not alone in this course of development. The Agricultural Revolution was driven by the human use of tools, but also necessitated the development of new tools. The Industrial Revolution did the same, and now the Data and Information Revolution followed the pattern of being created by technology and necessitating the invention of new technologies. Technology is shaping the world, and the world is shaping technology. The world uses technology in new ways, and new processes are dictating the direction of technology.

Whilst economics may have fallen short by attempting to analyse the behaviour of non-rational actors in a logical, rational way, behavioural economics and technology could provide a more plausible method. Because tech companies acquire data first and then find a commercial application for it, there is no data that does not have potential value. However, there is some data that can be harder to acquire. It may not always be data than can be traced to a specific individual but rather a geographical or socio-economic group. Medical information, for example, which is anonymised, but may still hold useful clues about who, when and for how long people suffer from, say, flu.

The commercial value of data, from banks to pharmaceutical companies to supermarkets, is clear. It can target marketing and help businesses be more efficient by predicting consumer behaviour. For access, directly or indirectly, businesses pay tech companies well. Governments and public services, however, do not always have the funds to pay for the services they need to provide, let alone purchase the data insights they might need. Fortunately for all concerned, apart from, possibly, the public, the government does have access to something of value that is not cash.

The public sector has access to data that could otherwise be hard to acquire. Contemporary and historical data on health, tax, crime and education that might be of great value, especially because much of the other data had already been given up by users. Tech companies have money, resources, innovative ideas and a need for data to improve their user profiles. Governments make decisions and rules with significant commercial implications whilst also needing money and better ways to organise and plan.

Technology, like science, is apolitical; it has no ideology. The tech companies behind it are also apolitical, but for more pragmatic reasons. They go where the power and commercial advantage is. Like many businesses, especially media businesses, they support whoever is in power and in charge of regulating them. The political right tends not to regulate generally, believing in small-state, low tax, individual freedom. As a result, big tech companies tend to publicly lean towards the left, displaying a love of openness, equality and democracy for all. Behind the scenes, tech companies rarely believe in any ideology and work with, as well as fund, all sides.

Government outsourcing is not new. The public sector has always used private enterprise for their expertise, to do something they should not, could not, or could not be seen to do. From mercenary soldiers and privateers to banks, construction companies and cleaners to consultants, paying outsiders with public money was commonplace.

The concept of outsourcing was radically overhauled during the late 1970s. It became less about temporarily bolstering resources in a time of need or accessing specific experts for specific projects. Outsourcing became a tool of the public sector to save money. The theory was that the private sector, made efficient by competition, could provide services better and therefore cheaper than the public sector. They were invited to make bids on everything from school meals to street cleaning to transport with the lowest bid invariably winning the contract. Outsourcing also led to a raft of regulations attempting to prevent corruption and nepotism as public cash flowed to private companies and individuals.

The tech companies, especially Google (along with its parent company Alphabet) and Amazon, have grown well beyond their original objectives of peerless search and retail services. They have sought out new applications for their vast financial and personnel resources. For their hard-tested research and development methods and techniques for generating new ideas. These companies have become heavily involved in the future of transport and logistics, medicine, hardware, entertainment, and the provision of internet access and infrastructure. These interests all share a common thread:

the acquisition of data.

Google not only knows what people search for but, via its Android operating system, it knows everything from when and where people buy their coffee (via Android Pay) to their location throughout the day, their health to what news they read. Amazon, as well as selling almost everything and delivering it, have developed a burgeoning entertainment business (Amazon Prime), and a giant data farm in the shape of its Amazon Web Services (AWS) business. AWS underpins the internet operation of companies and organisations around the world, including those that would nominally be seen as direct competitors, such as Netflix. AWS deal in so much data that in some cases the internet itself cannot deal with it so they physically ship it via storage units on trucks.

As well as the typical commercial operations, many of these subsidiaries and diversifications come together to enable tech companies to also act as large public sector outsourcing companies. All that data held by the companies is as useful to those seeking to manage public services as it is to those trying to sell goods and services. Furthermore, the tech companies are so wealthy and employ so many highly skilled, highly remunerated people that governments will show a great deal of flexibility to tempt them to set up camp in their constituency. If financial inducements are insufficient to see a company invest in an area, access to data, or a unique opportunity to harvest data others cannot, could win them over.

In Toronto, a company called Sidewalk Labs were given the contract to redevelop the long-neglected Quayside area of the city. On the surface, no different to any number of private developers winning a contract to regenerate an overlooked part of town. Except that Sidewalk Labs are part of Alphabet and unlike most urban and property developers their intention was not just to build some shiny new shops and offices. They will embed sensors in street furniture and buildings to harvest data on who is walking past, on air quality, data usage, transport and more. Sidewalk Labs may make some money from renting offices or selling refurbished buildings, but Alphabet will gain a much more valuable, longer-term asset; a dataset that is both unusual and huge. What has the City of Toronto lost? Nothing. Much needed work will be been done, jobs will be provided, a drain on their city budget will be removed, and they will have a short-term financial windfall from the sale. So far so unsurprising, except the City of Toronto did not know they had sold the development rights to Alphabet, having dealt with a third party.

Why would Alphabet not wish the City to know its data harvesting plans? What, ultimately, have Toronto's guardians signed its people up for? Alphabet may be using Toronto as a testing ground for a new form of urban development. Connected, intelligent, quick to construct, 'smart' and importantly, predictive. Flows of people, their demands and activities, the income and expenditure, this data combined with other services will provide new, more comprehensive profiles. In another project, Sidewalk Labs also replaced dozen of New York City phone booths with Wi-Fi kiosks and are working with the US Department of Transport on traffic, parking and public transport patterns.

In Colorado, a tract of land near Denver airport has been turned over to Japanese electronics company Panasonic. They are creating a project, similar to that in Toronto, of connected, driverless, sensor-riddled streets. Called CityNow, this $72m scheme is also working alongside the Department of Transport to build a stretch of road that creates a safer, more data-driven environment for driverless vehicles. The City of Denver has also embraced Panasonic, under the guise of improving citizens lives and lower financial burdens. Whilst effective in both regards, the cost is data that is not the City's to give. This is Panasonic's second experiment in urban planning after Fujisawa Sustainable Smart Town, 30 miles west of Tokyo. They are building environments that stream data constantly.

Amazon, the world's largest retailer, were in the market for a new headquarters. Seattle, the company's original home city, offered them unprecedented tax breaks to stay. Some might view it as a bribe so that City Hall can point to a demonstrable success. To City officials, it was a case of winning investment through clever negotiation and financial innovation. The tax break sees City income taxes collected from Amazon employees based in the new headquarters given straight back to the company.

Seattle is essentially giving Amazon money so that they will invest in the City, on Amazon's terms. Aside from the fact that Amazon may simply find ways to pay their employees in such a way as to avoid local income taxes, government uses tax in a way that suits (mostly) the public need. It often allocates spending to unpopular and uncommercial parts of public life. Amazon, or any business, will spend where they see fit, what suits their commercial priorities. Even if they are attempting to be altruistic, they will often get it wrong by not have the wider, inclusive view that government does.

This is not the only example of Amazon operating in this way. The company essentially created a bidding war of tax breaks between US cities including Atlanta and Pittsburgh. Just as private institutions have done for the last 200 years Amazon have used the threat of moving their facilitates to one place or another to gain favourable regulation or tax circumstances. Money, though, is no longer the only incentive.

In the UK, the tax inspection and collection agency, the HMRC, stands accused of letting Amazon off the hook when it comes to paying its corporate tax. Arguments over how tech companies pay tax, by virtue of their non-geographical structures, are long, complex, controversial and well documented. In a secret recording made public in the course of a separate legal case, a senior HMRC figure is heard to say that "I've heard from the Treasury; the Treasury didn't want us to be too hard on Amazon." Furthermore, British minsters had been provided with evidence that Amazon were essentially evading VAT (a tax on the sale of non-essential items) and that the HMRC were not being particularly thorough in pursuing them. Two years after that exchange, HMRC confirmed a deal with Amazon to provide all of their data storage and cloud services; something previously, by legislation, that was supplied by a number of small, British companies. HMRC chose Amazon as the best, lowest bid to provide these outsourced services.

For many years politics has been accused of being too close to the finance sector. Some of this intimacy was due to shared educational and social backgrounds. Some of it was the reliance on banking and investment as a source of tax, jobs and expertise. There were, of course, less wholesome relationships as the sector used its wealth to influence regulation and decision-making.

Frequently politics recruited from the financial sector, and when people left politics, many would go to banks and professional services companies. Frequently senior ministers or civil servants were seen to leave politics and join such companies as advisors, consultants or non-executive directors. This became so commonplace throughout the late 20th century that it became referred to as the 'revolving door'. By the 2010s, however, tech companies had started to overtake finance as the destination for, and source of many government personnel, particularly in the US.

The administration of Barak Obama, the President many thought of as the first of the social media age, was particularly prone to exchanging staff with technology companies. The President himself was known to have lobbied European legislators over the tax and regulation the latter sought to impose on Google, Apple and Amazon. Many commentators believe that, in time this intimacy could be seen as Obama's single greatest lapse of judgement.

The US might be the most obvious crucible for this sort of blurring of tech and government, but the small Baltic state of Estonia, proportionately, leaves even America in its wake. Estonia, the birthplace of Skype, is frequently labelled the most connected country in Europe. The implosion of Soviet rule in the country effectively forced it to start its economy from scratch. In 1996 the Tiigrihüpe (Estonian for Tiger's Leap) project became government policy, a key part of which was getting all schools online and well supplied with new computers.

Within 20 years the Estonian constitution enshrined internet access as a right on a par with food and shelter. Digital and handwritten signatures now hold the same position in law - cutting a swath through bureaucracy but also digitising every legal document. Taxes are paid and politicians are elected online, there is free Wi-Fi everywhere in the capital city of Tallinn, and children aged seven are taught coding.

Estonia is now home to cutting edge enterprises from robots to bespoke medicine. A small nation of around 1.3 million people, its friction-free digital systems have become a testing ground for a more connected society. The government is not just supportive of, but actively encourages ever-greater innovation in all branches of the public and private sector. So much so that, knowing it cannot encourage physical immigration beyond its size and means, it offers digital citizenships. People from anywhere in the world can live in Estonia digitally (for a price), without ever setting foot in the country. They do business and make use of Estonian services as an individual and as an Estonian (although they pay their taxes wherever they owe them). To some it represents the dream of a borderless world that many internet utopians hoped for.

In many respects Estonia resembles California's early growth – free, open, pioneering, writing the rules as it goes. It was once on the geographical frontline of the Cold War. More recently their Western-leaning outlook has seen them once again become a battleground, only this time in a Russia versus the West cyber-war. This has in turn placed Estonia at the forefront of the digital security industry. It has pioneered the use of blockchain and encryption in many areas of government and business, making it one of the most digitally secure places on the planet. Essential, given that almost every type of public and personal record and interaction is now digital.

The flip side to all this innovation is the ever-increasing dependency of government and public services on privately owned (often partly or largely foreign-owned) tech companies. The interaction between government and tech makes any notion of the Washington revolving door look like a petty concern. The race to turn Estonia into a tech hub-cum-proving ground is making its government into a board of directors.

Estonia's economy has raced ahead of almost all of its former Soviet peers. It has incredibly low government debt, unemployment and levels of corruption. It has high levels of disparity between urban and rural income. Tallinn may look like the future, but it is one where inequality is ever starker and harder to tackle.

In government and in business, tech companies have uncovered a new and powerful role. Responsible for the platforms that deliver ever more news and information, tech companies have an unprecedented capacity to shape opinion. They are also the largely unaccountable gatekeepers to services vital to the public.

Friction is the behavioural term used to describe the human inclination to inaction. Tech companies deliver or highlight information to users based on their past online activities. That means users have to overcome a degree of friction in order to actively seek out information other than that which is assumed to be relevant or of interest to them. By potentially altering people's access to information, under the guise of helping them, tech companies wield incredible influence. Furthermore, even without hacking through firewalls, generating fake news, or creating bogus accounts, there is enough publicly available information online to, in the right context, hugely affect share prices and elections. By selecting certain details, and ignoring others, users can easily be persuaded that lies are facts and vice versa. Most will not overcome the friction involved to seek out corroborating or contradictory evidence.

Once the printed press held almost definitive sway over public opinion and access to news. It was a press that, despite wielding great power with relatively little accountability, could at least be identified; the source of a mistake or a lie could be acknowledged and dealt with. This is often not the case with digital media.

Rumours and half-truths, misunderstandings and deliberate lies wield huge influence over the twin pillars of the Western world – market economics and democracy. Both of these concepts are more fragile than they are often portrayed as being. They are human confections reliant solely on human trust and belief. It is no surprise that both business and government, neither of whom really understand the tech companies, are now so beholden to them.

Tech companies have started to look increasingly like the old East India and Dutch East India trading companies, albeit with a sleeker, friendlier, more human face. They are so close to government that they are becoming almost indistinguishable. They influence the laws that are supposed to govern them and protect the public. They control virtual monopolies on the sources of certain types of data. They write new rules in previously lawless (in this case digital) territories. They affect markets and commodities prices and make profit from not producing anything. Processing raw materials (data), harvested at ever-lower cost, to create something for export and sale to those that will pay higher and higher prices for it (advertisers). They take advantage of a new world that government knows little of and can do little to affect. They are pioneering but they are also being corrupted by a lack of meaningful oversight.

Unlike the old trading companies, however, it seems doubtful that the cost of running a country will make the tech companies unsustainable or vulnerable to external shocks. Moral outrage over the abuse and exploitation of slaves will not be something they have to confront. Their possessions will not be broken up and nationalised.

Tech companies have shown themselves to be remarkably resilient to the sort of problems that, traditionally, would finish off other businesses. The loss of personal data through accident or hacking is yet to sink any large company. Even the Cambridge Analytica scandal that saw Facebook, through poor governance (or deliberate deception, depending on your view) hand over large amounts of user data did not see the latter significantly damaged. Such is the public's dependence on these companies. Calls to regulate and open up these private institutions are dismissed as being too complex or as a potentially dangerous precedent for government interference in business.

So vast and powerful, so ingrained into every aspect of life are the tech companies that there are observers who believe that governments should treat those companies more like another country. Some consider that idea from the perspective of a secretive, untrustworthy rival nation to be wary of and spied upon (who know, perhaps this is already happening). Others suggest treating the tech companies as a peer, to be given a seat at the UN and other supranational bodies; incorporated into the global decision-making process. Those that indulge in foretelling the Biblical end of days, when a single government leads the world, are shifting their gaze from the UN to Google.

From this point, many people will start to predict that humanity is on the road to a horrifying, dystopian, technologically reliant future. A future of permanent surveillance, manipulation, authoritarian control and insidious rule by unaccountable mega-corporation. The sort of cautionary imagery more typically associated with science fiction.

21

AUTOMATED FUTURES

From *Metropolis* to *2001* to *Bladerunner*, a central, enduring and recurring theme of science fiction has been robots, androids and automata designed to emulate human behaviour. They have equally been a preoccupation of real-world science. Machines have been made to entertain, to prove humanity's dominance over nature, to further understanding, to do what humans cannot (by virtue of greater strength or accuracy), or to do dull, dangerous and dirty jobs so humans do not have to.

The history of the robot goes back to the 3rd century CE with devices making use of hydraulics and later clockwork, and by the 17th and 18th centuries human-like machines were being built. The advent of electricity brought more reliable and adaptable robots but it was not until the 1950s and 60s that a second wave of automation, powered by microelectronics, significantly impacted on human employment. Just as the Industrial Revolution saw human labour replaced by machine labour, so commerce found a new efficiency saving in replacing humans even further. Robots became ever more advanced, able to do more tasks with more precision. The next barrier was removing or reducing the human dependency on programming or controlling the robot; finding a way to enable the robot to learn as a human learned.

In art, the robot has held a unique grasp on the imagination. Homer's *Iliad*, from 1200BCE, features mention of robot-like creations and the man-machine has been a regular feature of fiction ever since. There were non-mechanical creations like the Golum or Frankenstein's creature, human-like beings made to protect or to serve their creators, but that acted as a metaphor for human hubris. Robots in fiction were usually under the complete control of their human masters. The word 'robot' itself is derived from the Czech for forced labour. Popular robots in film from *Metropolis'* Maria to *The Day the Earth Stood Still's* Gort to *The Forbidden Planet's* Robbie acted at the whim of their biological masters.

As both fiction and technology developed, writers and artists started to question what would happen when robots misunderstood their programming or malfunctioned. This was often a metaphor for what might be in store for a modern society exploiting slaves and workers that might one day revolt. Real-world science, meanwhile, sought to understand the human brain and in doing so, artificially replicate its actions. The view that human behaviour could be rationalised and explained by science implied it could be reduced to the mathematics of computers. It made a new form of manmade intelligence seem feasible.

The possibility of humans having the machines they created turn on them was disturbing enough. The prospect of machines that combined greater physical strength with a greater intelligence was worse. Most terrifying and telling of all was the proposition that if a machine could replicate human thought, what about human emotion? Could one exist without the other, and what were the consequences? Could humans cease to be unique?

The film *2001* (as opposed to Clarke's novelisation) has been said to reduce its human characters to a banal, mundane level of discourse and emotion. By contrast, the notorious AI computer HAL9000 seems to be a character on a par with any person. In being more human than the humans, more deserving of the viewers' empathy, its attempts to kill become more visceral, and its ultimate demise more emotionally engaging. Clarke and Kubrick, deliberately or not, proposed an emotional convergence between humans and machines. Computers becoming progressively more human, although still far from human, whilst humans become more mechanical, although not quite computers. The film seems to suggest computers themselves might be a new form of life.

Intelligent robots - machines that could act independent of human direction - increasingly became both a focus for, but also background players in fiction. By the 1980s, with computer technology including basic robots much more commonplace, artists frequently used them to reflect on the nature of intelligence and consciousness. Again, science fiction both imitated and pushed real-world science as technologists tried to create thinking machines. Machines that could pass Alan Turing's test in which a human could not tell if they were interacting with another person or a computer.

Turing proposed the test in the 1950s as he considered whether a machine could think as a human would, rather than just work to a programme, processing inputs according to a set of instructions. In 1956 computer scientist John McCarthy (another Stanford alumnus) coined the term artificial intelligence. Ever since it has been the pre-eminent term to describe a particular form of machine learning; something that does not just approximate human responses to given inputs but actually modifies its own understanding as it 'experiences' the world.

The challenge of translating the real world into something a computer could 'understand' took a significant leap forward with digital computing. Tragically Turing died before he could see how digital technology might start to revolutionise the ways by which a machine could learn. By measuring and assigning numbers to sounds, images and language they could all be stored and processed by computers. Moore's Law also ensured, over time, ever greater precision of measurement; ever more data to be processed.

Processing data quicker than humans ceased to be a challenge for computers long ago; learning, however, was another matter. A computer can store the picture of a dog and identify subsequent pictures of the same, but it needs pictures of every type of dog, from every angle, in every pose to be able recognise any dog. Humans can see a single breed of dog only a few times and know that other breeds are still dogs.

Ever greater capacity for data and ever faster processing speeds can only go so far in replicating the human mind. Psychology, consciousness and emotion have to be digitised and turned into data. Data is nothing more than numbers; numbers that measure and represent almost anything. Information sees data in context, a number with a background, a unit of temperature or distance next to it. Knowledge organises the information, categorises it; the temperature of the human body or how far it is from London to New York. Understanding sees that knowledge interpreted; connections formed and acted upon according to other information. Wisdom is the product of experience, of failure and the repeated use of understanding in different circumstances.

AI is limited so far to somewhere between the information and understanding levels of this process, whilst humans work (mostly) at the wisdom level. What is generally viewed as AI is really only a set of incredibly complex, very rapidly processed algorithms – formulae that take an input or inputs and delivers an output. As processing power increases the more complex and numerous the algorithms, enabling something similar to AI; systems that appear to 'understand' the spoken word, or 'learn' a person's preferences, but it is not yet AI. It can store past requests, it can find patterns (of varying reliability), but it is not intelligent.

A NEW INTELLIGENCE

As, over time, the guidance of spiritual and religious advisors gave way to that of economists, so technologists in turn have come to overtake them. Each group in their own era and in their own way tried to predict the world and advised those in charge. The people looked to their leaders for certainty, unable to accept that certainty was the one thing no one could truly offer. Leaders had to reassure the people the best they could by relying on arcane, complex guidance few could reasonably question.

Religion's principal concern was morality to a point where it claimed a virtual monopoly on the issue. In time, morality sprung from the evolution of society – killing, theft and intolerance did not lead to productive societies whilst generosity and kindness did. Despite this, religion's ability to shape society, for good and ill, has nonetheless been spectacular. Economics has established a track record of successfully analysing capital and work. However, its over-reliance on rationality, mathematics and a desire to emulate science has been key to discrediting it or at least reigning in its influence.

Humans continually tried to find patterns, first in religious texts and scriptures, then in behaviours, ideas and numbers. The limitations of human intellect to find such patterns means it was succeeded by machine intellect some time ago. The technologists' greatest weapon in making a more rational world is artificial intelligence. In particular, AI's capacity to find bewilderingly complex patterns within almost incomprehensible quantities of data.

Economics, psychology and neuroscience have found, and continue to find patterns in human behaviour. More of those patterns will become evident as computing power improves - bearing in mind we are approaching the limits of Moore's Law and now need a new breakthrough such as quantum or organic computing. Unfortunately, some of those patterns will merely look like patterns, and some will be based on unreliable data.

AI, in common with new developments throughout human history, brings with it suspicion and fear. The fear of the unknown. The mass-collection of personal data from multiple sources is new and troublesome. But AI needs it in order to build up a vision of the world, to understand it and to make predictions. What AI reveals in that data can be disturbing for the fragile human psyche.

In China, a state more comfortable than most with surveillance of the individual, the government is piloting an

AI-based system that rates the credit-worthiness of citizens based on their data. Data gathered predominantly online – from shopping, social media interaction, time spent watching videos and so on. With this data it will decide, without human intervention, whether a person should be given credit. This is one of a number of stories that are seen by many as the thin end of an extremely worrying wedge. How long before access to credit become access to services or to overseas travel visas?

In Chicago, the police department have trialled a system that analyses crime data. It takes information on high-risk areas of the city, as well environmental data and data on individuals with past convictions for violence and brings them together. They try to intervene early with those who have social links to convicted offenders. They use GPS data from phones to assess if a high-risk person is entering a high-risk area and allocate resources accordingly. However, the system is prejudiced. The data on past crimes, offenders and sentencing input to the system is based on a history of targeting ethnic minorities, or those who live in certain parts of town, and of the very human prejudices of the public, police and judicial system. Any such system is only as good as the data its given, and in the US, potentially more than anywhere else in the world, a disproportionate number of black and Hispanic people have been through the justice system.

The parallel with Philip K Dick's *Minority Report* is obvious. But how much concern about this is due to humans fearing change and how much of it is genuinely, rationally justified? This type of policing could be more reliable than human intervention alone. It is dispassionate and logical. It works only on the evidence given to it (even if part of that is flawed historical data). Police have tried to pre-empt criminal activity for years by focusing on areas or individuals who fit certain profiles. The AI system will not always be right, but given there is evidence to suggest that human police officers can be affected by anything from the time until the end of their shift to their family circumstances at least AI will be consistent. Human biases are seen to be more extreme and more unpredictable than machine ones. They are also harder to account for and to correct over time.

Yet we are more comfortable with human biases because we too have them. We think the mistakes will even out and probably in our favour. We feel that a human can be made to see our perspective; to empathise with us.

There are years of debate due on whether society wants a system that dispassionately profiles potential criminals or credit risks and the role of human politicians in deciding how to programme such AIs (for example, should it consider ethnicity or socio-economic background). But these are early days and no one will rely solely on these AIs as they operate today. Machines will only be trusted when they too are part of a conversation on how to manage society.

The potential of AI to make independent, unbiased decisions is a source of hope as well as fear. AI views all data equally and can learn when to trust data and when to question it, as well as when to refer to an outside agency, typically humans. Much is made of who controls and designs AI and what their motives might be. As AI becomes increasingly commercial it should be considered how past tech pioneers started out well intentioned and believing they could predict and even control their creations. When released into the human world, however, the imperfect, unpredictable stakeholders, whether users, advertisers or investors, changed their creations.

If the gulf in resources and knowledge between government and Silicon Valley is so great as to make regulation all but impossible, perhaps the two groups will simply come together. Technology has the potential to understand humans, but not necessarily to make human-like decisions. It can measure but not always predict. In creating comprehensive profiles of people, rather than using them for commercial purposes, they could aid governmental and community discussion and decision-making.

The ability of AI and the technology around it to gather, compile and process vast amounts of data can help bridge the divide we see opening up in society; between leaders and the led, as well as amongst groups of individuals. AI can bring together opinions and voices, but in an intelligent, unbiased way; a way that helps leaders to make informed decisions. A way that will not be offended or provoked. Democracy, often criticised as the worse form of government apart from all the others (by Winston Churchill, but possibly by others before him and certainly by many after) and as the tyranny of the majority, has an obvious problem. Not every vote is equal. It is assumed that things average out, but how safe an assumption is that?

We are faced with a time, not too far away, where every interaction and experience, both online and real-world, generates data. In doing so, a person's data profile will say who they are, what they know, their background, job, income, opinions, habits, education, talents and skills. This knowledge can help create social cohesion. It can help schools create diverse classes and curriculums; help employers recruit diverse groups of employees. It can help authorities to plan communities and help businesses spring up around bringing people, and more importantly their experiences, together. It can ensure that a range of voices are incorporated, but it can also moderate the extremes by highlighting them as such. One person-one vote, the tyranny of the majority, would become one person-one voice; a voice weighted in importance according to who and what it represents. A public record of the knowledge that voice has; the contribution it has made elsewhere.

AI will certainly be able to achieve this - to assign an appropriate weight to a particular voice or voices. It could ensure a representative voice according to the question. By understanding how likely it is that a person has understood and considered a decision, it could measure the most amount of good for the most amount of people, whilst not ignoring everyone else.

Unlike opinion polls that take a sample representation and multiply it by the population it hopes to analyse, AI can handle entire nations of individuals like this. It can produce advice that truly takes everyone into account and produce a result that is not just the best for everyone, but nuanced according to their circumstances. In this, AI is a tool for decision-makers as well as wider society. It can reduce division; open up conversations, rebuild trust. It can expose people to the views and ideas of others whilst reassuring them theirs has been accounted for as well. Furthermore, it could encourage (or nudge) people to research the topic they are contributing to as to do so would increase the weight given to their voice.

This is a complex solution for complex times, but it might help steer the inevitable change in direction in such a way that everyone feels they have a part in it. And what of AI's ability to steer decisions on an individual level? Its capacity to deal rapidly with trillions of bits of data on billions of people means it can personalise. It can individually design online experiences, tailor exposure to information, but also help inform the design of the real world in which we live.

By applying the ideas of nudging, the behavioural science of making you fit a pattern, of making you more predictable, of subverting the human inclination to think short-term, AI is a potentially powerful tool. AI, with input from economics, psychology and neuroscience could make people predictable. By turning human experience and history into data and feeding it to AI, patterns in behaviour could be uncovered. New methods of nudging could be exposed. After astrology and economics a new predictive science could soon emerge. AI could prove to be the key to a real psychohistory by both the strength of its analysis, but also its ability to make people act more rationally. It could push into line those that do not fit the models. It could make them healthier, more economically prudent, but it could also turn them into more voracious consumers, more ideologically driven, depending on who controls the AI.

Silicon Valley's understanding of both addiction and 'me-ism' has created a possibly dangerous environment. A business model has immerged that has made a commodity of personal data (derived from an ever-increasing number of connected devices and services) and the psychological profiles that it generates. Ironically, the culture that persuades people they are unique is also the culture that is trying, and often proving, they are quite predictable and far from unique.

THE NEXT STEPS IN PREDICTION

Each generation of scientist has stood on the shoulders

of the last. They have taken the certainties of the past and tried to prove them wrong. Sometimes, they succeeded; other times, they simply furthered the original understanding or spun off in new directions. In making important discoveries an element of human creativity has been necessary. Creativity in imagining the far-future, extrapolating where science and technology might lead, how humans could act and societies evolve. As science has used creativity, so artists have applied an element of science to their creativity.

Artists are, by their nature, risk-takers. Their reputations are built on taking risks, by placing their hard work up for public judgement. The best artists are willing to gamble their reputations by continuing to innovate, to explore the unknown, to move away from the familiar and the popular. They are used to doing or saying things that challenge, that upset, that feel uncomfortable, and that attempt to be new, bold and different regardless of the repercussions. It is an attitude they share with the best scientists.

Shelly, Wells, Huxley, Orwell; they all creatively looked to the future. In doing so they produced great works of literature, but they have also been lionised for their foresight; for their predictions. They did not foresee many significant events, but then they did not set themselves up as predictors of technology or society. Their works matter more as commentaries on humanity; on our flaws, our ideologies, and our nature.

The works of these writers hold many important lessons for humanity, but they do not really serve as predictions, nor were they intended to. We foist our contemporary ideas on to these visions. Commentators of every age and inclination use such works for their own ends. They impose their interpretations, claim they understand what the long-dead author intended, and use that as a justification for saying we are on an inevitable slide towards our doom unless we act now.

Our own psychology, our own creativity, retrospectively fits works of fiction to our own experience. It is exciting to talk about Orwell warning us about the corrosive nature of digital mass-surveillance. We not only feel cleverer but also that our assessment must be right because an intelligent man 60 years ago said something vaguely similar. Orwell did not, could not, predict the gradual shift we experienced in the real world over that time; the changes of behaviour, the adaptation to living alongside technology. *Nineteen Eighty-Four's* surveillance state, likely based on the Soviet OGLU or NKVD secret police, held more of a warning about the East German Stasi (that would formally appear a year after the book's publication) as it would about CCTV in our high streets or the connected assistant in our homes. Likewise Huxley warned readers of the dangers of eugenics (an uncomfortably fashionable idea in the late 19th century that would go on to have terrifying consequences in the 1930s and 40s) more than any suggestion of foreseeing modern genetic engineering or cloning.

Many artistic visions have been inspired by the human desire to achieve the impossible. Flight, machines that think, space travel - these all had foundations in real science and were engaging ideas for fiction because the public wished that they were real. They wished strongly enough that their desire became a commercial impetus to pursue them in the real world. Young readers and filmgoers also grew up to be scientists and engineers as well, inspired to attempt to make those incredible visions a reality.

It might be reasonable to assume that the commercialisation of this sort of future thinking was the literature, film and television of science fiction. However, HG Wells suggested that the study of the future should be a science and an academic subject in its own right. Like economics, what was a philosophical concern became a commercial one. In the second half of the 20th century, future studies and futurology came into being. They combined science with social, political, economic and cultural studies along with features of management and strategic theory, leading some to suggest it might be a hybrid that is less than the sum of its parts.

First looked to by government and military planners, futurology was adopted by businesses to assist their planning, risk and decision-making processes. Technology in particular, both directly and its effects on society, brought with it a real and perceived increase in the rate of change. It meant that companies, who traditionally just had to understand their product and their customers, increasingly had to understand factors beyond their usual sphere of concern.

As with economic forecasting so with futurology: companies need to inspire confidence amongst investors, and part of that means reassuring investors that plans for the future of the company are robust. That they are prepared to weather any storm, no matter from where it might come. This saw companies employing futurologists; people charged with understanding aspects of the world that could affect a business, informing company strategy and by extension reassuring stakeholders.

The scenarios mapped out by futurologists would prove to company boards and shareholders that they had serious plans for the future; that they were prepared both for the threats and opportunities. The advice and strategy delivered by futurologists on what would happen when the oil ran out or when people no longer shopped on high streets proved companies were serious in building a long-term sustainable business. Far-fetched science fiction-style visions of a future where all cars were electric or everyone shopped via a small, flat piece of metal and plastic from their bathroom (or perhaps they did not shop at all with retailers predicting what consumers wanted) would not reassure anyone.

Like economics, futurology emulates certain elements of science, without being empirically a science. Like economics, futurology studies history, searching for trends and extrapolating patterns in past events (whether or not patterns really exist). Like economists, many futurologists or futurists served to feed Confirmation Biases by endorsing or validating the decisions of leaders, be they in politics or business. Futurology enables those in charge to point to 'evidence' that their decision was the best one given the information at the time.

Futurology, also like economics, has its own language, theories and models, often involving elements of management theory. As such it directly addresses strategies, thinking techniques, and strictly corporate concerns like product design, competitor analysis or customer service.

Unlike economics, futurology is more comfortable with change, wide-ranging and quite unexpected consequences, and chaos. Branches of futurology concern themselves with what the potential connections between events are and what the nature of their ultimate impact on the organisation or sector in question might be. They then assign a degree of probability to these events.

This use of probability, and the probability of one event happening which then affects the probability of another event happening and so on, naturally leads to some very complex mathematics and therefore a reliance on computer processing. Again, there is the temptation to see the presence of complex maths as synonymous with scientific precision. This in turn delivers a seemingly unbiased justification for whichever decision has been agreed upon, based on the futurologist's scenarios.

This methodology has been applied in business but adapted simply for government with a greater emphasis on political, environmental and social elements. Commercial concerns found parallels in public, national and international contexts. The analysis of business competitors was replaced by security service intelligence; customer service replaced by assessments of the impact on voter or popular opinion.

With a reliance on trends and the assumption that history shows the way it follows that, like economics, some futurologists get it right and some get it wrong. There are always those that successfully predicted big events, even if the world did not take them seriously. In the 1960s and 70s most futurologists did not suggest the widespread availability of home computing. A decade or so later even fewer accurately speculated on the effects of the smartphone. It is a common criticism of futurology that it either looks so close as to be of little use for any real planning or too far ahead to ever be meaningfully proved wrong. It also tends to deliver a range of possibilities and allows clients and audiences to follow their own preferred option.

In the years after the 2007-08 financial crisis demand for economists in business and government increased. As society deals with the Data Revolution the demand for futurists, for those offering a clearer vision through the apparently ever more crowded future has also increased.

Business and government stand to gain significantly by being well prepared and by making the right decisions. Equally, there is a risk of losing a great deal by making the wrong call. As such, there is money to be made in the prediction business. This has given rise to huge numbers of theories, models and disciplines within futurology. Those concerned about the future can call upon several techniques and models including trends analysis (trying to spot a pattern in what has already happened), scenario planning (based around sketching out plausible situations that will directly or indirectly affect a business or organisation), and morphological analysis (analysing all possible solutions to a complex problem).

Science has established that events like the movement of the planets and the stars as observed are predictable and fit relatively simple patterns. They obey physical laws and can be accurately described (in this instance by gravity and general relativity). Despite being looked to for guidance in human affairs, these celestial bodies have no actual impact in such matters. They continue in their movement by dint of natural law, not because humans see them or are somehow metaphysically connected to them.

Humans - creative, diverse, flawed, complex beings - introduce a random factor to events. Science can only go so far in accounting for their influence and predicting the outcome. Yet those charged (with or without consent) with ruling over groups of humans continue to seek certainty in the predictions they procure. A certainty born of past experience, whether it is the movement of share prices or how humans have responded to new technologies.

Futurology is just the latest discipline to try to predict

a world that has constantly defied attempts at trying to be predicted. Amongst all the complex language and modelling futurologists suggest that by examining, for example, the social, political and commercial impact of the train, we can predict current and future developments. The likely trajectory and effects of developments in the car, the airplane, the driverless-autonomous vehicle, space travel, teleportation and so on. In making such a prediction, and by offering selected evidence to support it, a government or business feels justified in making their plans. They can make the right provisions and investments in new technologies and skills to be prepared for, and even pioneer, the future.

In particle physics, Heisenberg's uncertainty principle states that there is a limit to how precisely two physical properties can be measured at the same time. For example, in accurately measuring a particle's position, its velocity cannot be precisely determined, and vice versa. In economics and futurology, the direction of travel, the rate of change, and the destination are all, it appears, impossible to predict with any degree of accuracy. Is there any reason to think that this will not always be the case?

Despite citing Shelley's warnings about playing God or Orwell's concern about unaccountable surveillance, society has not collapsed. It has adapted, learned, legislated, resisted, debated, and finally learned to get along with the new. Whether it is social, political and technological, the changes were never entirely good or entirely bad. The path they led humanity towards was as much governed by our own instincts as by some sense of fate; an inevitable, unwavering route towards a conclusive outcome. Change, neither wholly positive nor wholly negative, is the only thing that can reliably be predicted.

22

WHERE WE ARE NOW

In his book *The Better Angels of Our Nature*, psychologist and writer Stephen Pinker carefully outlines the general, long-term trajectory of humanity. It is almost uniformly positive and progressive. Human societies have, over time and without exception, become healthier, wealthier, more comfortable, less exposed to fatal illnesses, less like to die violently, better educated, and more empathetic. All the evidence suggests that this has been the broad, overall case for centuries.

Yet many believe we are worse off; that society, whatever society we know best, is deteriorating. That we live in increasingly violent, immoral, unhappy, divided, threatening times with no real, significant sign of improvement. Part of this may be due to ever more marketing subtly convincing us we are unhappy and insecure in order to sell us products. Part of it is certainly the human inability to view the world and its history in a wider context - as well as struggling to plan long-term we also struggle to look back long-term; to see the world in a broad, historical context. Some of it is our suspicion of change and some of it is simple self-pity. As responsible for our feelings of malaise as any of these is the unprecedented exposure to information, particularly negative information, that we are experiencing.

Once, humanity's only concern was what happened in our village because that was all we knew about. We moved to cities, learned to read and we started to worry about things we would never experience first-hand, although they might affect us indirectly. Perhaps just as self-defeating we worried about things we could not affect. Over time more people had access to more knowledge about the world, although it still required some time and effort - it meant dedicating time, effort and money to travelling or burying ourselves in books.

Today, however, we can learn (sometimes whether we want to or not) almost anything about anything, anyone or anywhere with just a few clicks. With the removal of almost all barriers to learning (if not understanding) comes an exposure to ever more stories of suffering and threat. The tragedies and the dramas that shock us are particularly noticeable. Humans became the superpower in the world in part by being able to spot dangers, communicate them to others, and learn from them. It enabled us to prepare for, and in many cases to master the threats we faced. Our brains continue seek out potential warnings, tending to remember them above the positive, reassuring information.

As well as communication, humans thrived and came to rule the Earth through adaptability, curiosity and creativity. Curiosity has seen humans find new lands and resources. It led to science and technology. Creativity enabled stories and lessons to be passed on through a variety of means, and its benefits have been obvious. These two facets of humanity, however, make us quite ill suited to a world where every bit of human knowledge and experience is potentially available. We overwhelm ourselves. We are acutely conscious of how much there is to learn and we feel under pressure to consume as much of it as possible.

Whether it is celebrity gossip or developments in politics or what friends are up to on holiday, there is a constant barrage of news alerts, updates, emails, and messages. Distractions masquerading as vital information. Each one demanding that we check them for fear of missing out; fear of not knowing something before others do. Each time we check, a small release of addictive dopamine. We sign up to apps and features that 'gamify' our lives; that set us targets, and with each target achieved, more dopamine, as well as more data on us and another thing to worry about.

As examined in behavioural economics, we feel financial losses more greatly than we do gains (our gain typically has to be almost double our loss in order to assuage our sense of feeling hard done by). In a similar way, we feel more keenly that which we are missing over that which we are acquiring. So it is that, knowing all the information is out there, feeling a degree of social pressure to know about as much of it as possible, but inevitably failing, is creating a uniquely modern form of stress. A feeling of being both left behind and overwhelmed at the same time.

If humans dominated the planet through curiosity, creativity and communication, the fourth ingredient of our success was change. Humans are endlessly, surprisingly adaptable. We live in all types of environments and societies. We have mastered or at least understood a great deal of nature and built systems and structures of ever-greater complexity.

We have always engendered change, and dealt with it, often in unexpected ways. But change used to be slow and gradual; much easier to adapt to and accept.

Along with exposing us to more information, technology has increased the rate of change in almost every area of our lives. The last decade has seen more change than the period between 1975 and 1985 (the start of the home computer era) or even 1810 and 1820 (the tail end of the Industrial Revolution).

In 1836 Samuel Morse revolutionised the world with the telegraph, enabling almost instant communication over great distances. In 1876 Alexander Graham Bell unveiled the telephone. Although it would be many years until the telephone was in widespread use, and a century until it was more-or-less in every Western home, it took forty years for one great innovation to give way to the next. In the forty years to today the average person has seen typewriters give way to wordprocessors and then computers, analogue to digital, the carphone to the smartphone; they have witnessed the internet grow to the point of ubiquity, and almost the entire lifespan of fax machines.

A big part of this is due to the influence of Moore's Law whereby what was possible ten years ago has increased in capacity or speed 60-fold (that is, if a computer could process 1Mb a second in 1975 in 1985 it could process 64MB a second and by 1995 it would be over 8,000MB per second). It also means cheaper, smaller and more computing power, so more of our world, from cars to fridges to front doors have become home to microprocessors and transmitters of data.

"I've come up with a set of rules that describe our reactions to technology:

1. Anything that is in the world when you're born is

normal and ordinary and is just a natural part of the way the world works.

2. Anything that's invented between when you're 15 and 35 is new and exciting and revolutionary and you can probably get a career in it.

3. Anything invented after you're 35 is against the natural order of things." – **Douglas Adams**

Humanity may have gotten this far by successfully adapting but we now feel, as a species, that we are not being allowed the time to adapt. We are not given the chance to acclimatise to and understand new technologies and their possibilities. Nor are we permitted to come to terms with what technology is facilitating – the changes in society, knowledge, and culture.

Communication has always removed barriers and created understanding. That was true of those communicating with neighbouring villages or journeying to foreign lands, those assigned pen pals at school or those in online discussion forums. Mass, open communication, easy to access and where almost anyone has a voice has built communities that were almost inconceivable a decade or so ago. Their voices have added to the maelstrom of information demanding our attention, as well as engendering change.

New groups, previously encumbered by geography or social mores, have banded together. They express their pain or frustration at being ignored and now demand recognition and consideration. Technology has created a non-geographical world where people gather around causes, traits or attributes rather than the arbitrary concept of nation state or locale. Those of a similar disposition or shared experience can find confidence and security. This is both positive and negative. This new digital environment has given those who have questioned their own gender identity consolation and advice. It has supported and even won justice for the abused. But it has also enabled racists and bigots to reinforce their prejudices amongst themself. It has spread rumour and uninformed opinion and disguised it as fact.

Another consequence of this sudden flourishing of previously hidden voices has been to confuse many outside of those groups. What many thought were certainties in life, things largely unchanged for generations - concepts of politics, nationhood, language, even humanity itself - have been challenged by new, very visible groups brought together online.

Being made aware of previously unheard ideas, opinions and experiences is rarely bad for society or individuals; indeed, it is usually quite the opposite. However, for many, the rate at which these new voices have emerged and influenced the world has just added to the feeling of being overwhelmed. They feel pressure to learn and adopt a new set of rules. They are told they do not really understand the world in which they live. This is as applicable to the average person as it is to political and business leaders - they are, after all, just people too. More than that, they are people responsible for and beholden to others, some of whom will be amongst these groups who have found their voices.

These new conversations, between leaders and the led, have seen the former questioned more; their words, actions and decisions scrutinised. The velvet curtain has been pulled back and their all-too-human frailties displayed. Leaders are no long special; no longer uniquely qualified to make life-affecting decisions on our behalf. They are one of us.

Infuriatingly for leaders this has led to a demand that

they act more like everyone else, whilst at the same time being better than everyone else. They must be moral, wise, effective, responsible, open and fair. They must account equally for all possible voices for whom they are responsible, and even those they are not. We demand more from our leaders than any previous generation, and we are able to check up on them more than any previous generation. This has led to a crisis of leadership, creating a class of risk-adverse, technocratic leaders (who are invariably over-reliant on data and historical patterns to justify their decisions) combined with a group of intensely private, unaccountable ones. If only leaders could lead a more rational population.

The rapid pace of change and the information overload has also seen an increase in mental health problems. Paradoxically it has also seen an increase in mental health awareness. The anonymity of the online world has allowed people to share their experiences and by doing so support each other and understand they are not wrong or freakish. They have found solidarity and made society more aware; research and treatment have improved and become more commonplace. It has gone some way to persuading the ignorant that it is they that are wrong.

Being overwhelmed is now such a familiar feeling that along with the serious aspects of mental illness it has spawned a somewhat more exploitative commercial sector. Key to this industry are pseudo-psychological concepts like 'being present' and mindfulness. The latter in particular has inspired a growth in commercial activity from apps to books to YouTube channels - all making money from the notion of 'me time' and by exploiting those that feel overwhelmed. They are focused around attempts to filter out, albeit for short periods, the notifications, emails, and updates that increasingly plague many modern lives. There are even apps to help you to stop using your connected devices.

This represents the ongoing commercialisation of the self. The notion that the individual is special, fragile and perfectly imperfect has become almost cult-like. It is a notion the tech companies are commercially driven to sustain, and technically capable of perpetuating. It allows them to gather more data under the pretext of personalising their service and filtering out the undesirable information on the users' behalf.

The phenomenon (real and imagined) of overwhelmingly rapid change means many people now fear change, regardless of its nature, and seek, at least in principle, to return to simpler times. A time before the Data Revolution, before the internet. It is a simplistic, and obviously forlorn hope but one that has been seized upon in particular by politicians, but also by businesses. Everything from nationalist nostalgia for a world that never existed to travel experiences where phones are banned to slow food and artisan manufacturing has gained traction in recent years. People tire of their phones and computers, of the hectic pace of change, of the constant churn of the new, whilst remaining addicted to novelty and social media.

In commerce, many bigger businesses feel threatened by the rate of change, and are concerned about their ability to keep up. Small, disruptive startups with their agility and flexible business plans seem much better equipped and so big companies buy them up in the hope they might absorb their innovative, reactive nature. They pay fleets of advisors and consultants to help them react better and prepare for this new world. In this, the new era of rapid change has just created a new sector of the economy.

Some branches of business, particularly marketing, as well as politics have embraced the rate of change and sought to exploit it. Seeing people overwhelmed they have added to or magnified that feeling. They have nourished a general sense of insecurity and uncertainty. They have used the fact that people feel overwhelmed; that they have no time to research and reflect on decisions. They appeal to emotions, to irrational, impulsive, System One brains to win people over to their side.

Liberal democracy as practised in the capitalist world relies on a combination of irrational and rational; on emotions and logic. But as the logical side of politics, the side that manages and makes utilitarian decisions, diminishes in appeal what is left is the emotional. Politicians now sell a vision of a golden post-war era - some even preferring the Cold War world, or even an era of empire. An imagined time of certainty and strong leadership. Some leaders even sell the vision of an era found only in the pages of religious texts. All of which is dangerous given how much rests on the logical side of running, serving and maintaining a country filled with irrational human beings.

Still, with a feeling of being overwhelmed by change and an onslaught of opinions, information, services, products and ideas, it is no wonder some yearn for a nostalgic vision (that almost certainly never existed) of a simpler world. We have seen many elections and referenda fought out between rational, logical argument and those advocating a baseless vision of an old fashioned, simpler life bounded only by national boundaries and clear, binary decisions. It has been the latter – the emotional, the irrational - that has invariably won. In truth, those victories were more likely than many wanted to believe. However, like the economists warning of a crash, there was often little to be gained at the time from raising the alarm.

It has been said that at each election politicians usually fight the last election, referring to the fact they have tended to respond to the concerns of the public the last time they met them en masse. Politics is slow to change.

At the human, individual level, the pace of change has made traditional politics seem largely irrelevant. Partisan groups with old-fashioned, fixed ideas of left and right are slow, outpaced by the changes and the people principally behind them; the tech companies. People no longer see any contradiction in being both altruistic and socially conscious with being personally ambitious. They own their lives, their work, their skills and use them all selfishly and selflessly amongst groups of their choosing.

With the old political forces and ideas now apparently out of date, people turn to communities, group of like-minded people, to solve problems or to air grievances. If geography is not dead, it is at least increasingly irrelevant, further pushing traditional politics towards crisis. Groups have started to turn inwards, isolating themselves. They have found security and reassurance in online worlds where everyone agrees with them. They see conventional politics as inconvenient, threatening, and even irrelevant. Government, however, remains essential to the operation of a country. It may be necessary to separate government and politics.

Whilst communities have their benefits and have achieved great things, they have also developed a serious shortcoming. They engender a lack of trust, not within themselves, but without. Rather than debate or adapt they dismiss anything that challenges their beliefs and chosen narratives. Even what constitutes a fact has become a matter of dispute. Nothing is independent; everything is biased. Trust in politics, trust in science, trust in experts has all been eroded. At the same time trust has become a commodity. Its value, how trusted someone or something is, like financial markets, is subject to rumour, speculation, personal agendas, and the whims of irrational humans.

Traditionally it was for politics and religion to bring people together. Trade demanded trust, government and religion supplied it through shared laws and observances. Faith used morality and divine judgement over all people to ensure honesty. Most political philosophies, or at least the moderate versions, centred around a form of utilitarianism. Whilst coming from the left or the right in terms of the process, the goal was broadly the most amount of good for the most amount of people, with as few people as possible left out. The aim was to find a compromise everyone could live with. If people are dismissing politics, what can hope to bring them together? Religion has either faded in importance or become a source of division in much of the world. Business certainly cannot be trusted (it tends to use its money to break the rules it finds inconvenient) and the media (traditional and digital) are too biased by ownership, agendas and advertising to hold them to honest, independent account.

In the years after the Cold War, the notion that 'the West had won' resisted any serious challenges. Politics throughout the West leant right-wards, unquestioningly embracing the free and decreasingly regulated market. Even traditionally left-wing voices chose to erode their own ideological core. As ideological argument faded from prominence and outsourcing took over, government became about management; politics about management styles and strategies.

One result of this shift was that many of the poorer people they traditionally represented felt abandoned. Reinforced by the Confirmation Biases of boardrooms and government cabinets around the world politicians convinced most of the population it was inevitable. When this led to a banking crisis, almost two decades of one-sided argument meant there was a vacuum of ideas and of leadership. There was no one left to represent those that suffered the most as governments adopted austerity policies, companies downsized and outsourced to maintain profits, and those responsible walked away to invest afresh.

Where once politics and its notions of ideology mirrored religion, it has come to more closely resemble business. Politicians are managers; parties are brands; voters are consumers. As technology has come to dominate business, so it has come to dominate politics; as a tool but also an operational model. Tech companies have come to represent the acme of innovation, success and strategy. Tech companies understand their users and their employees in a way no government could ever hope to understand voters. They have an influence no political ideology could hope to have. Little wonder politics looks so enviously towards them and little wonder tech companies have governments pleading for their insight and assistance.

Business and business figures, meanwhile, have taken on areas more traditionally associated with government. Through research, outsourcing, commercialising previously non-commercial issues or philanthropic projects they display a human side. In fact, they get involved in what matters to them personally or commercially. Education, healthcare and development aid are as such more publicly acceptable or commercially useful than projects to help ex-gang members or prevent sexually transmitted diseases. Only government is able to take a broader, more cohesive and dispassionate view of these problems.

Unfortunately for politics, people look to their leaders for more than just managing. They seek certainty and reassurance. At the same time they demand openness and immediate, effective action. Yet the people themselves are the greatest source of uncertainty and inconsistency, leaving government unable to offer certainty because of the very people demanding it.

LIVING WITH UNCERTAINTY

Psychologically humans crave certainty. It seems only logical to us that somewhere there is a pattern. Surely the world cannot be random. Randomness unnerves us. For so long humans seemed masters of all they surveyed. They understood everything because all they needed to understand was that which was in their immediate, first-hand experience. Then along came science to change all that. It introduced uncertainty. Worse, it embraced uncertainty. It also brought about new technologies which further undermined humanity's assumed dominance of the world.

It is perhaps no surprise that the idea of a science that can shape and accurately predict human affairs is an attractive one. Mystics, priests and economists all aimed to do this. The latter has come closest and, along with psychology and neuroscience laid the foundation for what is to come. Technology has brought these subjects together. It has created communities but also reinforced a sense of the individual and in doing so it has harvested bewildering amounts of personal data. It has harvested this data, uncovered new behavioural patterns and realised new processes. Could this be the route to certainty and prediction?

Once we could rely on governments to serve us, companies to employ us, neighbours to support us. The online world, combined with real world problems some of which have technology at their core, has engendered a widespread breakdown of trust. We no longer believe our leaders to be better informed or wiser than us. We no longer trust businesses to play by the rules we have to play by. They in turn can no longer rely on us humans to speak with a united, rational voice. Trust, the cornerstone of our trading, government and social lives, has been eroded from all sides. People no longer trust those who disagree with their views, but equally politics and business do not trust the people to make sensible, considered decisions.

Politicians have long suspected people of being irrational; unable to judge what is best for themselves. Now, they have data to support that suspicion. Social media has highlighted public hypocrisy and inconsistency, whilst leaving politicians exposed and unable to defend themselves. As politics has become more like commerce, it has applied the same marketing and communication techniques. The online profiles of voters, rather than being used to tailor commercial messages and filter out 'irrelevant' ones are applied to politics. The problem being that influencing a consumer's choice of products is relatively unimportant. Influencing their politics requires affecting not just the adverts they see but the news and information they consume. It is not just about what brand of trainers or beer or shampoo they buy, it is about their worldview. Furthermore, their choice of political party or their endorsing of particular ideas can have a more profound effect on wider society as well as on them as individuals.

Tech companies offer a service to help users that feel overwhelmed by the size and scale of the internet. The companies filter out what, based on their profiles, users will want to see. The individual's role in seeking out information, or even in asserting what they might be interested in, is vastly reduced. Applying this to the commercial domain means, in theory, navigating the bewildering array of goods and services is a lot easier. Applying it in the non-commercial world means users are only exposed to a very narrow range of ideas and opinions; a selection of themes that has been algorithmically curated for them. Such restrictions lead to a difficulty in distinguishing fact from opinion. The application of psychology and a lack of regulation means news and adverts merge and are strategically placed to provoke particular reactions. All of this is new and mysterious to many users, and in particular to business and politics. Leaders feel they no longer understand those they are responsible for and neither group trusts each other.

Trust relies on a combination of the rational and irrational. Emotionally we must connect; logically we must understand the facts, as well as the difference between a fact and an opinion. The less contact we have with those from whom we differ, in whatever way, the less trust there will be. It does not matter if that is those that lead businesses or government or those that simply live or work elsewhere. Trust can only be rebuilt by open, inclusive conversations.

Believing that overall, as Stephen Pinker observes, humanity is actually on a course of positive advancement can inspire complacency. There are, however, some things we have to accept and some areas where we have to focus our efforts. We must accept change and uncertainty; the unpredictability that it brings. We must accept that humans are excellent at adapting to change and our main concern at the moment is the speed not the nature of that change. We must also accept that we have the facilities and access to the information to productively steer some of that change. It is beholden upon us to use that opportunity wisely. Humanity thrived through building communities. Restricting or fragmenting communities will only hold us back.

Technology, with its ability to connect people, might offer a solution. But technology is going to have to do more than just passively facilitate conversations. It does that already and there is rightly the fear that we will just end up continuing to talk only to those who we believe are like us. Connecting more people will not necessarily change that. The Confirmation Biases will remain, possibly even increasing in influence. The power of technology, especially AI, to analyse data needs to combine with the human desire to collaborate and to create that has been our species' greatest advantage so far.

If people do not trust others to do what is right - to at least consider opinions and experiences that are not their own - that might not be their fault. If they only come into contact with people similar to themselves online, and also in their office, their street or their school, that means they do not see the lives of others. They will not understand or empathise with them. These are the problems technology must address. Technology is a tool to solve problems, but it is necessary for humans to first understand those problems.

Society is confusing change for deterioration. We are

confusing rapid change for chronic deterioration. We fear change and long for certainty. So much so we tend to make ourselves more stressed by railing against the inevitable change. Technologists tend to solve the problems technology brings by adding more technology. Technologists, however, are only human, and will tend to create technologies in their own image, solving the problems they see and remaining ignorant of the ones they have never experienced. Indeed, their imperfect solutions often create yet more problems requiring more solutions - which is good for business but bad for society.

As with so many problems facing the world, open collaboration between technology, society and politics could hold the answer. No one area can be left to deal with these challenges, and this is where government's managerial role can come in. Government, which is morally responsible for all of its citizens and uniquely placed to see all the problems from above. Government needs to move beyond just creating laws that are applied and obeyed inconsistently. They must put aside the remnants of ideology and the constant strategising for power.

Government must bring together differing voices in an unbiased, independent manner. It should be above squabbling and manipulation, corruption and ego. If it cannot, it should make use of technology to achieve its goals of management, good judgement and inclusivity. Government must, however, be allowed to do this by those who put them in place, that entrust them with power, and to whom they should be answerable – the public. If humans hold their leaders in contempt, they can expect the same in return (at least in private). Citizens need to take their share of responsibility; understand the part they have to play in feeding the petty arguments and evasion that plagues politics. If government and the people fail to steer society and commerce in a positive way, business, specifically tech business, will take on the role as politicians lose interest in the job.

23

TECHNOLOGY AND THE FUTURE OF PREDICTION

Governments who are managers. Businesses who seek to get involved without fully understanding the nature of the problem. Politics and ideology no longer seen as the best systems for running the world. A society feeling overwhelmed by vast amounts of information and rapid change. Individuals who are convinced they are important, unique and deserve to be heard in each and every debate regardless of whether they are qualified. Tech companies building ever more detailed (but potentially flawed) profiles of their users, the world they inhabit and how they interact with it. Technologies that offer an open, unbiased and accountable way of recording and analysing decisions. The means to contribute to decisions, to guide government as individuals and groups both equally, yet with favour to those with the most to contribute. A technology married to science and social sciences that can both predict and make more predictable. A human desire for certainty – a desire to know what the future holds.

Asimov's psychohistory looks to many like a terrible curse. Humans reduced to predictable bystanders in their own lives. Governed by something that is not human but was created by humans that took on a life of its own. Asimov himself was ambivalent about the nature of his fictional Foundation society. Rather than making a moral judgement he considered human nature and the new societies we build in the image of previous ones. It is this inclination that makes human progress predictable through the new science of psychohistory.

Would having our futures mapped out and guided for us by a high-functioning intelligence applying a new science be all that bad? Could the everyday continue to be under human control, whilst the big picture, long-term plans we humans are so poor at comprehending is taken out of our hands?

It is to the advantage of government as well as tech companies, and business more generally, to make humans more predictable. Whether that is what they will buy, the health services they will require, or the policies they will agree with. For the first time in human history, it may be possible to make people behave in a determined manner. A new science with a connected, data saturated environment and AI at its core; a new way to understand and to manipulate humans.

As Aldous Huxley observed, we may be moving towards a world where individuals can only exist in exile from comfort and community – a technological exile, removed from the influence of the digital which predicts our needs. Those who opt out of the digital domain, thereby retaining ownership of their data and refusing to be part of the predictive model, would be punished. Not by anything so archaic as banishment from society as such, but by excluding themselves from the comforts of the modern, connected world, and perhaps, from the services managed by government via digital platforms.

At the heart of *Brave New World* lies a question: whether it is better to live in a good hell or a bad heaven. A choice between living in blissful (or at least comfortable) ignorance or knowing discomfort. The digital world where you can get lost, only hear positive things, be the person you want to be, and in exchange, hand over all your data. Or the real world; messy, awkward, sometimes painful and unpleasant, and always unpredictable.

As government and tech companies become closer, as the former comes to rely more on the latter, citizens become consumers. In the name of progress we are expected to digitally monitor and pay for utilities (gas, water, electric), to access information, to pay taxes, to maintain friendships. Remove ourselves from the digital domain and we stop ourselves living a 'normal' life, like those free-thinkers of *Brave New World*. The nudge – keep living within the digital domain so that you can be monitored; so that you can be usefully added into the predictive model.

Nietzsche saw the power of a society in its ability to stop a person being who they could be; achieving what they could achieve. Restricting access to goods and services to only those engaged in the digital leaves almost no choice. Nietzsche's superman was a thought experiment that could not live in the real world – he would need to trade, communicate and rely on the expertise of others. The digital world demands that we take part, play our role and at the same time places limitations on us.

There is an often-quoted exchange between a US official, part of Nixon's entourage on his visit to Beijing in 1972, and the Chinese premier Zhou Enlai. Asked for his thoughts on the

French Revolution 200 years earlier, he is said to have remarked: "it is too early to say". Whilst the veracity of this quote has been disputed, it is cited so often because it neatly encapsulates the drawback of trying to assess events that are, in the truest sense, chaotic – riven with unpredictable, long-term, indirect consequences.

Revolutions, be they political or socio-technological, take generations to reveal themselves. To resolve the interplay between different aspects, their geographical and cultural spread, the utterly unpredictable, chaotic consequences they produce. One thing has been proved repeatedly to be true about revolutions, from the Agricultural to the Russian to the Data – humans adapt. It may take time and it may be painful, but we ultimately live with and shape these disruptive events, individually and collectively, and very occasionally, under the leadership of key figures.

Humans fear change, yet we have consistently proven ourselves to be incredible at adapting to it. We fear uncertainty, we have expended huge amounts of energy and resources trying to prepare for any eventuality, yet it frequently makes little difference. Human consciousness, the desire to question and to feel, has cursed us with frustration at our own limitations. So much so that we sometimes deny those limitations even exist. We are condemned not to understand some things, and certainly never to understand everything. Our intelligence is limited and flawed, but also incredible because of those limits and the way we work around them.

Ludwig Wittgenstein commented extensively on what humans can know, what we need to assume in order to progress, and the limitations in knowledge placed on us by language. Our use of language to communicate with others and ourselves (our internal monologue), means we can only achieve so much. The human mind is incredible, but not always in the way we might expect. Its ability to see patterns is incredibly limited. We can only focus on one element or trace out one pattern at a time. Partly by using past experience to fill in the blanks our brains fool us into thinking that we are taking in everything around us, but we are not. Where the mind is incredible is in its ability to very quickly switch focus from one thing to another, to sample rapidly and to read in detail.

On the other hand, even basic AI is incredibly pattern-driven, far superior to any human in its ability as well as capacity to analyse, but it is also inflexible. The conclusions it draws are based on vast amounts of data, and a small degree of assumption. Humans, by contrast, can draw conclusions based on relatively little data and huge assumptions. Just like recognising many breeds of dog despite having only seen one or two.

Work on AI has essentially been little different to all previous forms of software programming, by codifying knowledge and inputting it, transferring it from the human world to the machine world. The problem being that humans cannot fully explain everything that they know. Not just complex, uniquely human emotions, but everyday activities, especially those relating to physical motion and perception. It is why we demonstrate rather than explain processes that are actually quite complex. The thinker Michael Polanyi gave his name to Polanyi's Paradox to describe how much of human learning is not conscious but passed on through often non-verbal cues. Summarised as 'we can know more than we can tell' it is why 'teaching' a machine is so difficult. Also, as far as AI is concerned, we can only teach in a human way, which is, if not useless, extremely limiting when teaching a non-human.

We have no better way to deal with this new technology. Supervised learning is the name given to the process most people associate with AI today. It involves thousands, sometimes millions of examples of something being fed into a computer, and the computer being 'taught' what they are so that they might recognise them again in the future. For example, voice recognition software, which is given thousands of examples of words and syllables being spoken in various accents and intonations. Having been so programmed, it then 'knows' what word it 'hears' probably getting it right even if it is said in an accent not previously encountered by the machine. This seems incredibly complex to us and requires millions of hours of programming and inputting to work, and yet still we find such systems incredibly limited in the real world.

This type of data-heavy instruction has seen AI employed by Amazon to predict stock control, by hospitals in cancer diagnosis, and by JP Morgan to review commercial loans. Information that would have taken hundreds of dedicated workers thousands of hours to sift through can now be done in a few seconds by very powerful computers. This looks like intelligence, but it is still some way off emulating organic intelligence.

The next leap forward will occur if and when a new form of machine learning develops; one that does not need petabytes of data fed into it. One that can 'see' one breed of dog and recognise all others also as dogs. Just as humans can. This is AGI (Artificial General Intelligence) and predictions vary as to whether it is possible at all, and if it is, by when.

AI is a hotly debated topic with many ethical complications. Regulation and ownership are key issues, with concerns over the replacement of human workers in similar ways to those expressed during the Industrial Revolution. Amongst the other debates is whether an intelligent machine would have rights similar to those bestowed upon other forms or life, or even humans.

"God created man in his own image." Ever since that idea was first proposed, probably somewhere around 500BCE, humans have seen themselves as the highest form of life (at least on this planet). The one best suited to the world. Writers and visionaries have imagined humans becoming gods by creating their own beings in their own likeness, both physically and mentally. We have a narrow view of the world and can only envisage intelligence in our own image.

We know that the way the human brain works is very different to how a computer, even the most powerful and complex computer, works. Yet we come back to an idea of a process of thinking and learning akin to our own. Despite the fact that computers do not (and need not) share our experiences, our language, or our physical connection to the world. We are flawed and limited but uniquely human. For example, our eyes can only sense certain frequencies of light, and suffer from myopia and colour blindness, or in some cases do not work at all. This means we understand the world in a particular way, and often in a way we cannot explain. It shapes our minds, allows us to imagine and create and guess and explore what we cannot see. True AGI, should it be possible, will have its own way of learning, of 'sensing' and gathering information, and of processing information, perhaps even of expressing itself, that we humans cannot. It may be better in some respects, worse in others, but it will not be the same. It might understand us, but it will not replace like with like.

Wittgenstein also famously said that if a lion could speak, we could not understand it. It would no longer be a lion, by its nature, it would be something else; something shaped by other experiences that would set it apart from other lions. Equally if a lion could speak, it would not speak any language, express any idea or worldview that a human could comprehend. So for AGI, which would cease to be the technology humans had created and become something else.

Currently AI is still in its adolescence. Much of what we take to be artificial intelligence is actually a cleverly designed algorithm applied to very large amounts of data via impressive amounts of processing power. AGI will learn without human interference; without being told what a set of data represents. Algorithms, even very complex ones, simply process data and provide a particular, albeit sometimes revealing, result.

Even in its nascent form AI has already established a good, if not always outstanding predictive track record. In healthcare it has found patterns in cancer diagnoses. There are banks who use AI to study historical market data and make investment decisions and insurance companies who use it to assess risk - sometimes wholly reliant on AI, sometimes combining AI with human decision-making. AI has demonstrated an ability to predict the movement of crowds at sporting events, whether students will drop out of college, even a person's sexuality with somewhere around a 90% accuracy. Whilst that sounds impressive, 10% could represent a huge number of people. Plus, if human leaders are reluctant to argue with the data-driven, more-often-right-than-wrong AI, the implications could be very serious.

The perfectly reasonable assumption is more information, more complex systems, and more application in the real world will only improve AI's predictive ability, with or without the leap to AGI. Although there may well be limits to what is and is not predictable, and those limits need to be identified, the application of AI needs to involve everyone, otherwise the data is biased and the results unfair. Currently much of this work is based on the data available either through government and academic systems or from those groups that engage extensively in social media and internet search. That leaves a large number of people unaccounted for should someone be creating a predictive model.

There is, of course, a paradox in tech companies persuading people they are unique in order to get them to engage with their platforms in order to harvest data from them and use that to build predictive models of their behaviour. It perhaps reveals the comfort we find in the irrational and the contradictory. But it is surprising just how much of society and of human behaviour is predictable, or has become predictable.

Technology is also changing our brains and how we think. Attention spans, the ability to remember facts, concepts and feelings of pleasure and loss are, some believe, being irrevocably altered on a physiological level. But does it matter? Are these changes just part of the process? Neither good nor bad - what we lose in one place we gain in another. Our affinity for remembering facts replaced by a skill in understanding how the internet orders and searches information. Are we losing what makes humans special? Or were we never really that special in the first place?

Whenever there is great change, some fight and some acquiesce. Some see an opportunity to profit, others to improve the world, but overall, we adapt. The Luddites of the Industrial Revolution are often mistakenly assumed to have been against the machine age simply by virtue of being anti-progress; of being somehow pathologically nostalgic. In fact, they felt passionately that industrialisation was a way to circumvent labour practices and that it would destroy hard-learned crafts and skills. Attacks on machines and factories saw military action and the suppression and arrest of Luddites. The British Parliament, populated as it was by many with financial or political stakes in factories, passed laws making the damaging of machines a criminal offence. Although defeated, the Luddite's ultimately played a role in the formation of trade unions and the development of workers' rights. It took decades for humans and machines to live alongside each other, and along the way they shaped each other and changed society for both good and ill.

AI is the latest advance in automation, intended to reduce the workers' share of dull, dangerous or dirty work. At the same time it is designed to remove humans and the costs they entail, enabling instead investment in machines which drives up profits for the owners. In the long term, the social trajectory industrialisation has produced is positive with millions of people enjoying a longer, healthier, more prosperous life. As has been demonstrated, we struggle with the long-term, whilst in the short-term we see only threat, disruption and discomfort. The Industrial Revolution created a new form of urbanised working class as semi-skilled and unskilled workers moved from farm to city. The unintended consequences were illness, social disorder, political turmoil, even war that took decades to recover from, but recover, and indeed thrive, is what societies did.

Globalisation then superseded industrialisation. Factories closed, the jobs they provided were moved thousands of miles away to less developed economies where costs (including labour and regulatory costs) were lower. In turn the economies of those towns and cities, that had largely grown up around one or two employers or industries, collapsed. Towns across the US and Western Europe became hollowed-out shells of addiction, mental illness, crime, joblessness and despair. Those individuals that could leave to work somewhere more prosperous often did so, leaving those that could not escape to cope with what was left.

An automated, AI-led, technological revolution could offer a route to recovery. People will no longer need to live in prosperous towns and cities where space is limited, and public finances are over-burdened. They will be freed from work, or at least partially freed, as many jobs still require some degree of human oversight or input. They will have more time to learn more skills. AI will free people. As the likes of Oscar Wilde and Matthew Arnold suggested, humanity's salvation lies in the freedom of the individual to pursue their innate vocation. To improve themselves without concern for economics. They should enjoy and indeed produce art for art's sake, not to prove their intellect, to further their place in society, or to earn money. That, to those idealists, is what freedom looks like.

For over a century people have hoped that a functional society could come from enabling everyone to pursue that which they are best suited to, rather than chaining them to desks or counters or machines. This utopian vision requires a more equal sharing of capital, and plays to a somewhat privileged, unrealistic view of the world. Marx insisted that this sort of freedom could only come about when the workers owned the capital; the means of production. But with no workers, who will profit and who will suffer?

One widely discussed option is the so-called robot tax; a tax on any business employing AI or automation to do a job humans could do or have done previously. But for how long will you be able to say that a job is a human job? Eventually robots will be the primary source not just of manual labour but a great deal of intellectual work as well. AI is already much better at many of the process-driven tasks humans currently fill their time with. AI will only continue to be quicker, more robust and less costly than humans. So who will differentiate between what is robot work for robots or where robots are supplanting humans? An alternative may be, rather than a financial tax, a legal duty of care to former and never-will-be employees; to support their continued learning and skills development. This would at least deliver the advantage of businesses helping people to earn, and in doing so help them to buy the goods and services the businesses provide. Business can only gain from helping to maintain a healthy (in all respects) society.

Concerns about AI and technology, whoever they are from, should not be written off as the anti-progressive, lunatic diatribe of those who are too slow, old or ignorant to adapt. We should worry about those feeling marginalised or excluded (directly and indirectly) from the technological wonders that so excite some parts of the world. We should listen to those who resist the steamroller of Silicon Valley's PR, marketing and lobbying industries. The fears of many sceptics need to be examined and applied, not brushed aside. That way a robust and inclusive future lies. Those with a stake in the Data Revolution should not be allowed to crush all opposition, no matter how persuasively they do it. Like the Luddites, opposition voices need to be involved in the decisions that shape a world that could well be on the brink of something remarkable or terrifying.

24

STAYING UNPREDICTABLE

The great strength of the internet, its greatest attraction to us, is its ability to treat us all as individuals. To give us what we want almost without us having to ask. It encourages us to believe we are unique and uniquely talented, interesting, and important. It tells us that our voices, our creativity should be shared with the world. In sharing our voices we are giving tech companies ever more data about us. Data with which to predict our behaviours and, potentially, to make us even more predictable.

We do not like the idea of being predictable. We are secretly, perhaps unconsciously, proud that our unpredictable natures have forced economists and others to come up with new models and new ways of looking at the world. We like the idea that technologists have seen their products fail or looked on confused as the public used them in completely unforeseen ways. We like the idea that we are unique and different.

Humans at once demand more certainty in the world, yet hate the idea that we ourselves might be predictable. Others are predictable and easily manipulated; we are too clever, too special. In some ways we are uniquely unpredictable, but we may also need to accept that in many ways we are indeed very predictable. It is a case of understanding how and under what circumstances we are predictable, and how that might be of benefit.

Machines have proven to be better than humans at everything from lifting heavy objects to calculating hard sums. They have been less adept in areas of creativity and judgement. As such, the priority should be to identify the areas where predicting humans is both possible and useful, not just remunerative. That has the potential to help us. To create a society that embraces data and technology beyond business goals. To free us and to help government to manage a demanding, complex world of individuals in a transparent, accountable society.

The workers, whether they are in fields or offices, factories or shops, have been the foundation of every revolution. They have either inspired and built it, or been irrevocably changed by it. Socialism - as its core, the freedom to lead a life unfettered by toil and drudgery - can be traced back to Plato, Aristotle and Epicurus as well as early Islam. The principle has frequently been corrupted when put into practice, but its possibility has engaged the imagination for centuries.

As AI and the automation of both manual and intellectual work has come ever closer to reality, so too has the idea that huge parts of the population could be made redundant. Not only builders or factory workers or shop assistants but also doctors and lawyers; professions many assumed would remain immune from automation. Whilst the skill and learning of these jobs might still be needed, the world may need fewer of them. If those roles do survive, they will be much changed and require quite different skills.

There have been two responses to this scenario: one is that society will edge closer to form of socialism. Towards a system whereby people are freed from the pressures and obligations of work and salaries to pursue their personal goals whilst still maintaining a high standard of living and efficient governance. The other response is a dystopian prediction of a fragmented, barely functioning society. A world of redundant, poor, ill-educated masses that are powerless against a niche of isolated, dynastic, technologically-enriched oligarchs.

The twin threat and promise of AI is ostensibly similar to that of the Industrial Revolution. Like that revolution the reality will invariably be somewhere in the middle dictated entirely by the human affinity for adaptation. The Industrial Revolution did not lead, long-term at least, to mass unemployment. The fragmentation of society was gradual, and more of a shift than a comprehensive degradation. It led to different types of jobs, never imagined a generation or two previously. It led to new, cheaper products, workers' rights, better housing, disposable income and leisure time, healthcare and mass education. It also led to social and psychological problems, precarious economies and the dominance of commerce. It is possible we will see some of these phenomena increase, disappear or change completely as AI and automation grow in importance.

Although history does not always repeat itself, the shift from industrial, manufacturing economies in the West, to service-based, intellectual economies does hold one cautionary example for a world of AI-based automation. Managed poorly, as the shift was in the UK from the late-1970s through to the 80s, the long-term results are severe social division and economic fragility. Managed properly, humans can adapt to the new conditions, shape them, own a stake in them, and prosper.

If AI holds the key to the next great social revolution, it will take time. Neither the technology nor our response to it will emerge overnight. Both will evolve, as individuals, groups and governments shape and are shaped by advances in technology, commerce and society. And because our response will be unpredictable it makes preparing for it almost impossible. Trying to regulate or defend against undesirable applications of AI will be undermined by the pace of change and by our unpredictable, irrational response.

There is a belief that AI has the potential to both completely understand, and eventually replicate, human behaviour. This would only be possible were AI to essentially be human. To have the same growth and experiences. AI will not mirror the human mind; it will be a new type of intelligence, different to anything seen before. More efficient than our human intelligence, probably, but not the same; flawed, but in a different way to us.

Can a new intelligence ever accurately understand and model the human mind? A mind that follows patterns sometimes, but not always (and under what circumstances it does follow a pattern is governed by quite irrational factors). A mind made up of a bewildering array of influences over a lifetime. A mind that incorporate genetics, physiognomy and psychology. That is shaped by unpredictable experiences, environment, social interaction and imperfect senses. A mind that is wise and foolish at the same time and that is defined by its limitations as much as its potential. If such a technology can understand the human experience, an inherited and learned state; then truly AI has the potential to surpass and replace humans.

In reality, something else is much more likely: the appearance of a new intelligence, something that may well require the coining of a new word, that we work alongside. Something that we could communicate with, but not necessarily empathise with, and vice versa. A new machine (although again, that word may cease to be appropriate) that, like every tool since the first rock or stick was sharpened by humanoid ancestors over 3 million years ago, we use and that changes us.

It is this concept of AI that humans would need to adapt to. It could very well replace great swaths of the working population, but that population will adapt as it has done so many times before. We may delegate tasks to AI, but we will also work alongside it, perhaps as a tool in the way we currently use computers, or perhaps more like a colleague. AI will assist but not necessarily dictate our decision-making. Just as humans need a variety of skills, perspectives and experiences in order to make good decisions, so AI will join that diversity as a new type of voice.

In the aftermath of the financial crisis of 2007-08, questions were asked about what nearly destroyed the global banking system. Whilst the realities of how banks had come to view risk and their attitude to other people's money were central, what had led to those views, and worse, was a defective corporate culture. One of the most corrosive elements of that culture was a demonstrable, ingrained insularity.

For years those in charge of organisations had recruited and promoted in their own image; a combination of deliberate policy (hiring from a shallow, familiar pool), unconscious bias (not realising what they were doing) and arrogance (assuming the best people for the job were those like them). We all have embedded within us an inclination to surround ourselves with similar people, similar opinions, and information that back up our beliefs. It is the duty of those with power to ensure this does not happen to them. It may be the role of technology to ensure this sort of behaviour does not lead to poor decisions or fractured societies.

By having decision-making bodies made up almost exclusively of people of similar backgrounds, education and outlooks, banks failed to embrace other ways of thinking and understood only how their world, not the real world worked. This lack of diversity and openness, and the Confirmation Biases it enabled to go unchecked, was not just a cause of the crisis, but also inspired shortcomings in dealing with it.

In the period after the crisis it became clear that these problems were not limited to the financial world. Businesses of all types, frequently through whistle-blowers and social media, were revealed to be home to destructive, exclusive cultures. Cultures that those outside looked at with bewilderment and frequently disgust. Environments where senior figures were not or could not be challenged, where those of different backgrounds were excluded (perhaps because they could not be trusted to toe the line), where insecurity manifested itself as hubris and where the outside world rarely encroached.

This type of insularity, and the complacency it engendered, was thought by many as infecting experts of all types, including economists, and particularly the political classes - perhaps the group who have had their status most diminished. This contributed in large part to what has come to be seen as the twin shocks of popular democratic discontent in recent years – the 2016 US Presidential campaign and the UK-EU membership referendum of the same year. Those in politics, media and associated sectors were found to be too insular and failed to understand those whose lives they affected.

All of this has seen diversity become not just a moral issue (which it undoubtedly is) but a commercial one too. A question of effectiveness, not just civilisation. Some organisations have slowly realised that whilst it makes for good publicity to have a senior team reflect the wider world, it actually makes for a more productive and resilient business. If all the voices contributing to a decision essentially sounded the same, how could you be sure a decision was robust - that it would serve everyone it needed to? Where would new ideas come from? The answer is not just diversity in the sense it is most often assumed – of gender or race (although that comes into it) – but of socio-economic background, culture, education and age. A genuine diversity of thought and experience.

The biases, the insularity, the uniformity of appearance and experience, the resistance to change; they have been traits of boardrooms and corridors of power for decades, for centuries. Every so often a brave, indomitable pioneer would break through; become a role model and change their world. Barriers are broken down, expectations redefined, and by this, society progresses. It realises the value of more and more of its members. In the years since the financial crisis, in the age of the new, user-orientated internet, with increased transparency, with more shared experiences, this change too has intensified.

With all of this, a generation (not necessarily by age, but by outlook) of business and political leaders are looking increasingly irrelevant. Their way of working, of relying on data and patterns to justify their decisions, and humans to carry them out, is being replaced. Replaced by a more human, empathetic leadership style. A style that withstands the modern demands for transparency and social media scrutiny. A style that values collaboration, openness, and the ability to make abstract connections. A style that cannot be better carried out by a machine. When this leadership model is combined with, and takes into consideration, machine intelligence it will make for a powerful, and even positive, business or government.

Whilst the diversity of human voices becomes, slowly, a reality, the acceptance of an AI voice as equal to human voices is a long way off. It is also the potential source of serious problems. But it may also be important not to think of AI as necessarily humanoid in appearance, thought and speech.

The second problem can be understood when looking at the nature of the information any decision-maker is faced with. Increasingly data is seen as the most important element to reaching any decision. With enough data, the theory goes, the right course will become obvious. Asking 'what does the data tell us' means those making the decision have a get-out clause if it goes wrong. It means they do not have to make a decision themselves.

"Doubt is not a pleasant condition, but certainty is absurd." - **Voltaire**

The internet has made everyone believe they are an economist, a historian, sociologist, politician and lawyer (as well as doctor, mechanic, chef, computer programmer...) This misplaced confidence in one's own knowledge and ability to use freely available information has created a newfound difficulty in being aware of how little we actually do know. It has become harder to separate truth from fiction, opinion from fact, news from propaganda, honest reportage from malicious lies. We have neither the time (we are overwhelmed) nor inclination (we know what we agree with and what we do not) to really look into these matters. Yet we do not trust those whose job it is to understand these things for us. We do not trust them because we do not think they have our best interests at heart. Ironically, we ourselves frequently do not have our own best interests at heart either.

This distrust, even antipathy towards experts, as well as the human irrational view of what benefits us long-term, is the other big factor in those two recent democratic shocks – the Presidential election and the UK-EU referendum. These events matter because they concisely demonstrate the difficulty in modern decision-making. The error in relying on ideology to govern a public that wants to wrestle control from its traditional leaders but will not study all of the pertinent facts. Regardless of which side of the argument someone may be on, it is undeniable that those events have divided their respective countries in a way few issues have before and represent a real change in society.

An intelligent human has an idea of their intellectual limitations. They understand that, as science has demonstrated, they cannot be wholly certain about anything. Even if there are historical patterns that seem to suggest they can be. They know when it is best to keep options open, when to hedge bets, and when to defer to a better-qualified person. They realise that a diversity of opinions will lead to the best decision.

This may be where AI could help in the future, and potentially where it should be directed. Its ability to dispassionately analyse human behaviour, past events and patterns, whilst also calculating probability means that it may have a valuable contribution to make. From the most mundane, everyday consumer decision to those national and international ones affecting millions. The best results for any course of action, strategy or decision is one that is formed through diverse opinions and ideas instead of relying on one or two, no matter how experienced or well-suited. AI has the potential to account for and balance out human biases, and even to recognise when the system itself is being relied upon too heavily.

Science fiction has for some time placed artificial intelligence amongst humans, not as servants but as peers and advisors. In Ridley Scott's film *Alien*, all spacecraft carry a science officer in the same way that many expeditions in the 17th and 18th centuries carried a scientist. Like their forebears, AI's role was somewhat questionable as they combined the pursuit of knowledge with less admirable, colonial or commercial ambitions. In Scott's vision, that scientist was always an android. A more rational, exacting version of a human. It was viewed as equal to anyone in the crew, but the captain (a human) still led; they commanded, but like any good commander they embraced the knowledge of their crew. Where expertise was required, just as they would refer to the (human) engineer, they would refer to the (non-human) scientist. Where a vote was required, the humans and non-humans had equal say.

AI's greatest contribution will be as part of the solution, not being relied upon as the whole solution. Tech companies are, regrettably, still largely reliant on white, English speaking, middle class men to steer their products and strategy. As such the AI they create will hold, at core, their unconscious biases. The machines will tackle the problems their makers see as more important. They will learn from the data that may or may not be wholly representative. This must be tackled first before their version of AI gains an unstoppable momentum. This is the responsibility of both government and society, who have as much power as consumers as they do as voters.

We fear change, and no one will have experienced a change on the potential scale of AI. We are wary of trusting AI, and not just because 200 years of science fiction has convinced us that 'playing God' is dangerous. We will need to accept though that as flawed as AI might be, it is certainly no more flawed than human decision-making. Slowly, through assistants on our phones and in our homes, we are embracing AI, but we also need to understand it in order to shape it. We cannot leave it to an increasingly remote tech elite.

It is to be hoped that AI will free us from decisions that we are ill suited to make. Decisions that require too much time and information to make them properly. We do not always understand our own self-interest, especially in the long-term. AI will be able to understand our self-interest better than we can and so we should come to trust it. It can free us from feeling overwhelmed, but are we willing to open ourselves up to it in order that it can learn enough about us to make decisions on our behalves? And where is the limit? Are we happy to let AI choose our meals and choice of film, but not our job?

Work continues to occupy our concerns with AI; specifically, its potential to replace us. If we do not want to be replaced, we must adapt. Commentators implore political leaders to teach coding or Mandarin Chinese to modern school children, and even to those beyond compulsory education and in the workforce. However, there are already automated, AI systems that can code. Foreign languages are already translated by AI, and their skill will only improve and become ever more portable. At the time of writing, Waverly Labs have already developed a prototype earpiece that translates spoken languages into the wearer's ear.

The result of this means an onus on government to develop a new form of education. Learning throughout life, yes, but an ongoing development of our creative, curious, abstract abilities. Even assuming that AI, eventually, realises a form of creativity, it will not be the same as human creativity because AI experience will not be human experience. Education needs to move away from learning facts, processes, even languages that machines can (or very soon will) do better. Understanding how to work with AI, understanding how decisions are best made, appreciating what diversity means are all far more important than learning French verbs, US Presidents or how to calculate the volume of a cone.

Regardless of the hope of idealists like Wilde and Arnold, their hopes and visions are unlikely to be a sound template for global progress. It may be a sad truth that whilst some today may learn to tan leather or carve stone, the mass need for those skills died decades, even centuries ago. Those that can do them well may charge a premium for their work to those that want products made by human hands, but it is often as much a source of pleasure for the worker as for the consumer. It is no basis for a modern economy. In the same way speakers of languages and those that can read maps may become little more than hobbyists in the future.

Humans need to focus on their unique ability to solve problems, create, form connections, learn from history, and have original ideas. Whilst abstract and potentially very hard to measure, this will be much more useful and less likely to become outdated. Technology, through applications like virtual reality as well as AI, may even find new and more effective ways to teach these skills, using more than taking notes from a lecture, and tailoring learning to a student's individual needs.

Once again, Silicon Valley has led the way, albeit in a limited, insular way. Some of the tech elite have, like Gatling and Nobel before them, realised their originally altruistic intentions have gone awry. Whether through dubious business leadership or the simple, inexorable corruption of unregulated capitalism their work has become weaponised. In an attempt to fix some of the more corrosive aspects of their legacies they have started to work on solutions, trying to reduce the addictive, manipulative applications of their algorithms, as well as the more underhanded data-harvesting techniques.

At the same time, the Silicon Valley elite have started to home school their children or send them to expensive institutions with new ideas about education and that limit access to technology. They encourage technologically independent, creative thinking. In this they seem to want to lead the way for others. But deliberately or not they are buying their children an advantage in the same way generations of plutocrats before them have. Their children will be better equipped, more immune, better prepared than most others of their generation to work in an age where being able to think irrationally and act unpredictably is a much sought-after skill.

Although probably not what either Adam Smith or Milton Friedman ever had in mind, capitalism, under the influence of digital technology, has seen their notions of Human Capital become ever more real. If we accept that, by and large, each individual is now their own means of production, their own factory in which each of us must invest, develop and improve, the metaphor can be extended. In the same way that the Industrial Revolution and everything that came after has seen the owners of the capital (in this case, us) make innovative, judicious use of automation and machines, we should do the same. In the era of the Data Revolution that means improving our own efficiency through AI. Trusting machines to make quicker and more robust decisions on the logical, the rational, pattern-based aspects of life. Freeing humans to focus on their uniquely human irrational, creative, curious traits. Traits that, like any skill, will die out without continual use.

Capitalism has a history of forcing innovation on to the people. New jobs, new ways of living. In that, AI is just another phenomenon for humans to adapt to. Meanwhile in China a whole country is seeking to redefine capitalism as dictated for 200 years by the West. It is challenging received notions of everything modern capitalism is centred around, from working practices to regulation to concepts of collaboration and competition. These forces could, gradually, change our everyday lives in ways similar to the Agricultural and Industrial Revolutions.

Humans will adapt; as jobs are replaced, new jobs will emerge; jobs alongside AI, working in concert with AI, interpreting AI. Just as the internet has created new jobs and new ways of doing old jobs. Professional forecasters as late as the 1980s did not foresee the internet, so there was no chance they would predict a world of user experience designers, social media PR agents and people who spent their days researching online. Who knows what a world where AI aids big-picture, long-term decisions and takes over our tedious small-scale ones, holds for humans?

Were the worst to happen and millions (some say as many as 50%) of jobs became fully automated, what then? One solution to a world of decreasing employment is a universal basic income – a guaranteed income for everyone. However many economists suggest that this is a less effective use of money than a properly funded, well managed welfare state with proper healthcare, (lifelong) education and other strong public services. In a world where the nature of work has shifted we may need economists, experts and students in the nature of work, to help unpick the path ahead.

No technological advance is entirely good or entirely bad; only the responses to it. Were AI and automation to have a devastating effect on employment (which would be gradual), it would invariably be used to solve it. The capacity of AI to assess huge amounts of data means a welfare system and services that could measure each citizen and assess them according to their needs and abilities. A truly fair, individually tailored system could be developed. As in government so in business. Employers who could understand the needs, circumstances and skills of their individual employees, enabling them to allocate work and pay fairly. People would feel less marginalised whilst also being appropriately supported and encouraged.

Looking once more to Asimov's *Foundation*, is it conceivable that with AI's advances a new social science will come to the fore? Could AI be the element that helps gather together human expertise in economics, politics, history, anthropology, psychology, physics and see everything differently? Could this lead to a new economics; a behavioural, technological science that can predict and correct human mistakes before they happen? Making larger-scale events more predictable, but resisting making individuals more predictable in their daily lives. Saving humans from themselves when they do not understand what is in their self-interest, but allowing them to engage in their own lives. Advising, not dictating. If that was the case, we all need to accept the argument for a better, more trustworthy form of government; a government less reliant on commerce.

A new political paradigm would be required. One that works with technology and not for it. One that uses it to make informed decisions for everyone's benefit. That can apply technology creatively. One free of irrational ideology, but also less in thrall to the irrational demands of citizens. Chiefly government must manage and stop measuring. Management means understanding those that you are managing and the goals for the organisation. You cannot manage if you are spending a significant part of your resources measuring what you are supposed to be managing.

Currently technology is being largely used to measure. It can take more measurements and process new targets. Technology could, however, offer the solution to the simplistic measurement that plagues politically driven government. It could assess development and improvement rather than the achievement of arbitrary numbers like arrests made or surgical procedures performed.

Nowhere is the hegemony of the number and target more pernicious than in education. Schools are almost exclusively measured on their ability to have children regurgitate facts and replicate learned processes. This is because in no area is it more difficult to measure improvement and success than in education. Facts are clean and measurable. They are right or wrong and if you get more right than wrong, you have achieved. This has been the way for over a century of universal education and it has barely changed. Successive governments have just tinkered with the substance of what is learned, rarely the nature of it.

Education is the key to future success, whether that success is defined as personal contentment or national productivity. Whether the aim is global concord and collaboration or simply continued economic growth, the focus will need to be away from that which technology comprehensively covers – the learning of facts - and towards more abstract skills.

Whilst some traditional subjects have the important consequence of teaching ways of thinking, and of course it will remain vital to be able to communicate verbally, many others need to be rethought. Education should produce those that understand the world and their responsibilities, but also freethinkers, schooled in the limitations of facts and the advantages of thinking creatively through learning logic, history, philosophy. In some respects a return not to the learning of classics but learning in a classical way; a rounded, progressive way. A way that questions and challenges, researches, and understands. That combines the creative with the ability to analyse; not learning facts but learning what they look like, how to test them and how to use them. That is open and can find connections between the apparently unconnected.

This is the case for future generations, but it is also true for us all. We all need to learn, relearn, or at least sharpen our curious and creative natures. We are predictable and will only become more so, but we need to hold on to and develop our irrational natures. We need to value that which makes us human and to improve those aspects. We need to guard the part of us that will always be unpredictable.

ACKNOWLEDGEMENTS AND THANKS

Special thanks first and foremost to Charlotte for all the support, patience, advice (most of which I took) and proofreading. She is the reason this book exists and I couldn't have written it without her.

Thanks to Sarah Smith for her design and typesetting work and guidance, and also thanks to Priya Lakhani of Century Tech for taking time to discuss this book generally as well as artificial intelligence specifically.

Any book that spans such a variety of topics as this one is bound to be informed and inspired by a myriad of sources. Facts have been taken and ideas have spun off from TV and radio programmes, articles and online videos too numerous to credit but particular mention should go to the BBC, The Guardian and Aeon.

Undoubtedly the single greatest resource of facts and ideas has been hearing from certain individuals talking about their specialist areas of research and expertise in person. This was made possible by JLA (**www.jla.co.uk**) who, as well as paying me a salary, afforded me the rare opportunity to speak to or hear from the following people, all of whom (unknowingly) in some way informed some part of this book: Dan Ariely, Rachel Botsman, Nick Chater, Cameron Colqhuoun, Tom Fletcher, Tom Friedman, Hannah Fry, Nicholas Gruen, Yuval Noah Harari, Daniel Hulme, Elizabeth Linder, Adam Rutherford, Yancey Strickler, and Sam Willis. This book is my attempt at bringing together ideas and insights from all of these people, and many others, and trying to find a common theme throughout human history.

Finally, the following books, whilst not specifically cited, should be acknowledged as having made an overall contribution to this book, its shape, as well as various facts:

Sapiens – A Brief History of Humankind by Yuval Noah Harari

A Brief History of the World by HG Wells

A History of Economics – The Past as the Present by JK Galbraith

NOTES AND REFERENCES

CHAPTER 1
Gascoigne, Bamber History of Astrology, HistoryWorld From 2001, ongoing.
www.historyworld.net/wrldhis/PlainTextHistories.asp?historyid=ac32

CHAPTER 3
Rowlatt, Justin. *Why do we value gold?* BBC World Service
(8 Dec 2013)
https://www.bbc.co.uk/news/magazine-25255957
Graeber, David. *Debt: The First 5,000 Years*, Melville House (2013)
Brezis, Elise S. The Oxford Encyclopaedia of Economic History, Oxford University Press (2003)
Selected pages from Investopedia, **www.investopedia.com**

CHAPTER 8
Bartlett, Robert C. *An Introduction to Hesiod's Works and Days*, The Review of Politics Vol. 68, No. 2 (Spring, 2006), pp. 177-205
Vance, Erik. *The Amazing Lives of Fan Li and Xi Shi, The Last Word On Nothing* (2015)
http://www.lastwordonnothing.com/2015/04/01/the-amazing-lives-of-fan-li-and-xi-shi/
Selected pages from the Adam Smith Institute **https://www.adamsmith.org/the-wealth-of-nations/**

CHAPTER 9
Burtt, Edwin Arthur. *The English philosophers from Bacon to Mill*, Random House (1997)

CHAPTER 10
Goodman, Peter S. *A Fresh Look at the Apostle of Free Markets*, New York Times

(April 2008)
https://www.nytimes.com/2008/04/13/weekinreview/13goodman.html
McCumber, John ed. Dresser, Sam. *America's hidden philosophy*, aeon (2017)
https://aeon.co/essays/how-cold-war-philosophy-permeates-us-society-to-this-day

CHAPTER 11

Seybold, Matt ed. Marina Benjamin. *Confidence Tricks* , aeon (2018)
https://aeon.co/essays/how-cold-war-philosophy-permeates-us-society-to-this-day
Keyes, John Maynard compiler Skidelsky, Robert. *The Essential Keynes*, Penguin (2015)

CHAPTER 12

Sullivan, J Courtney. *How Diamonds Became Forever*, New York Times (May 2013) **www.nytimes.com/2013/05/05/fashion/weddings/how-americans-learned-to-love-diamonds.html?pagewanted=1&_r=1**
Hrdy, Sarah Blaffer. *Mothers and Others - The Evolutionary Origins of Mutual Understanding*, Belknap Press (2011)

CHAPTER 13

Morgan, Daniel Patrick. *Astral Sciences in Early Imperial China: Observation, Sagehood and the Individual*, Cambridge University Press (2017)

CHAPTER 14

Why people don't tend to forecast recessions – The Economist (Buttonwood's notebook July 2016)
https://www.economist.com/buttonwoods-notebook/2016/07/21/why-people-dont-tend-to-forecast-recessions

Tetlock, Philip, & Gardner, Dan. *Superforecasting: The Art and Science of Prediction*, Random House (2016)
Chapter 15
Samson, Alan. *An Introduction to Behavioural Economics, Behavioural Economics*
https://www.behavioraleconomics.com/introduction-behavioral-economics/

Ariely, Dan. *The Upside of Irrationality: The Unexpected Benefits of Defying Logic*, Harper Perennial (2011)
Thayler, Richard H, & Sunstein, Cass R. *Nudge: Improving Decisions About Health*, Wealth and Happiness, Penguin (2009)
Kahneman, Daniel. *Thinking, Fast and Slow*, Penguin (2012)

NOTES AND REFERENCES

CHAPTER 16
Levinson, Marc. *The Box: How the Shipping Container Made the World Smaller and the World Economy Bigger*, Princeton University Press (2008)
Harford, Tim. *Fifty Things that Made the Modern Economy*, Little, Brown (2017)
Scaruffi, Piero. *A Timeline of Silicon Valley* **www.scaruffi.com/svhistory/silicon.html**
Bus, Natalia. *From hippies to Silicon Valley: the birth of California design lies in Sixties counterculture*, New Statesman (August 2017)
https://www.newstatesman.com/culture/art-design/2017/08/hippies-silicon-valley-birth-california-design-lies-sixties

CHAPTER 18
Warr, Philippa. *Facebook reveals your high IQ*, Wired (March 2013)
www.wired.co.uk/article/facebook-personality-predictions
Stix, Gary. *Animal Study Finds a Brain Circuit That Spurs Bullying*, Scientific America (June 2016) **https://blogs.scientificamerican.com/talking-back/animal-study-finds-a-brain-circuit-that-spurs-bullying/**
Dopamine, *Psychology Today* **https://www.psychologytoday.com/us/basics/dopamine**
Bergland, Christopher. *Neuroscience Suggests That We're All "Wired" for Addiction*, Psychology Today (August 2016)
https://www.psychologytoday.com/us/blog/the-athletes-way/201608/neuroscience-suggests-were-all-wired-addiction
Smith, Fran. *How Science Is Unlocking the Secrets of Addiction*, National Geographic (September 2017)
https://www.nationalgeographic.com/magazine/2017/09/the-addicted-brain/
The Science of Addiction, National Geographic (video produced by Mike Olcott)
https://www.nationalgeographic.com/magazine/2017/09/science-of-addiction/

CHAPTER 20
Zaretsky, Robert. *The Mytilenean Dialogue From 428 B.C. Explains Who Really Won the Trump-Clinton Debate*, **foreignpolicy.com** (September 2016)
https://foreignpolicy.com/2016/09/28/the-mytilenean-dialogue-from-428-bce-explains-who-really-won-the-trump-clinton-debate/

Walt, Vivienne. *Is This Tiny European Nation a Preview of Our Tech Future?* Fortune (April 2017)
http://fortune.com/2017/04/27/estonia-digital-life-tech-startups/

CHAPTER 21
Rosenblat, Alex, Kneese, Tamara & Boyd, Danah. *Predicting Human Behavior*, for Data & Society Research Institute (2014)
https://datasociety.net/pubs/2014-0317/PredictingHumanBehaviorPrimer.pdf

Brynjolfsson, Erik & McAfee, Andrew. *The Business of Artificial Intelligence*, Harvard Business Review
https://hbr.org/cover-story/2017/07/the-business-of-artificial-intelligence

CHAPTER 22

Glei, Jocelyn K. *Unsubscribe: How to Kill Email Anxiety, Avoid Distractions, and Get Real Work Done*, PublicAffairs (2016)

Botsman, Rachel. *Who Can You Trust?: How Technology Brought Us Together – and Why It Could Drive Us Apart*, Portfolio Penguin (2017)

INDEX

Agricultural Revolution, 11, 49, 58, 92, 135, 182, 191, 357
Alien (film), 426
Allen, Raymond, 282
Allende, Salvador, 356
Amazon (company), 277, 285, 300, 305-307, 312, 315, 359-364, 410,
American Civil War, 106, 264
Anchoring Effect, 243, 244
Antikythera mechanism, 18
Apollo (god), 19, 20
Apple (company), 271, 277, 285, 300, 305, 307, 309, 310, 345, 364
Arnold, Matthew, 415, 428
ARPANET, 270, 276
Arthashastra, 117
Asimov, Isaac, 336, 341, 342, 346, 406, 431
Atari, 277
Babson, Roger, 186
Babylonia, 15, 16, 31
Becker, Gary, 165, 166
Bengal Famine, 111
Bentham, Jeremy, 140, 141, 242
Berners-Lee, Tim, 276, 284
Black Death, 58, 59
Bodin, Jean, 122
Borgia (family), 41, 42
Boston Tea Party, 112
Brin, Sergey, 284, 306
Buddhism 25, 26, 273
California, 262-316, 365, 439
California Gold Rush, 263, 266, 316
Cambridge Analytica, 368
Canard, Nicolas-François, 139
Carnegie, Andrew, 148, 180, 204
Cassandra, 20, 232

Catmull, Ed, 345
China, 40, 53, 57-66, 78, 183, 194, 261, 347, 374, 430, 438
Christianity, 27-29, 42
Clarke, Arthur C, 336-340, 347, 371
Club of Rome, 174
Cold War, The, 159-178, 268, 272, 283-287, 295, 336, 343-346, 354, 356, 365, 395, 397, 438
Confucianism, 26, 60
Containerisation, 259-261
Cook, James, 79, 85
Cournot, Antoine Augustin, 139
Croesus (King of Lydia), 20
Darwin, Charles, 85, 96, 97, 189, 195
Dick, Philip K, 346, 375
Duchamp, Marcel, 309
Dunant, Jean-Henri, 146
Dunbar, Robin, 28
Dutch East India Company, 69, 36, 70, 71, 367
East India Company, 69, 109-112, 367
Einstein, Albert, 90, 213, 214
Elizabeth I, 67, 68
Facebook, 286, 287, 299-307, 312, 231, 368
Fairchild (company), 268-271, 277
Financial crisis 2007-08, 141, 207, 211, 214, 217, 234, 237, 244, 245, 384, 421
Fisher, Irving, 230, 242
FitzRoy, Robert, 85
Foundation Series, 341, 342, 406, 431
Franklin, Benjamin, 136

Freeman, Richard, **229**
Freud, Sigmund, **189, 196**
Friedman, Milton, **162-166, 429, 436**
Fukuyama, Francis, **283, 284**
Galiani, Ferdinando, **125**
Gallup, George, **290, 291**
Game Theory, **176, 177, 242**
Gatling, Richard, **80, 308, 429**
Gerety, Frances, **203**
Gigerenzer, Gert, **247**
Goldman Sachs, **209**
Google, **285, 287, 300-315, 359, 360, 364, 368**
Graeber, David, **52**
Harbou, Thea von, **343**
Harrison, John, **92**
Heinlein, Robert, **336**
Henry VIII, **40, 67**
Hesiod, **115, 166**
Hewlett, Bill & Packard, Dave, **266**
Hinduism, **26**
Hobbes, Thomas, **95, 123, 124, 128, 145**
Hohlenstein-Stadel lion-man, **15**
Hot/Cold Cognition, **246**
Huxley, Aldous, **332-335, 380, 381, 406**
IBM, **269, 277, 291, 337, 340**
Industrial Revolution, **95, 101-108, 127, 138, 142, 143, 157, 161, 189, 257-260, 265, 357, 369, 390, 411, 413, 414, 419, 429, 430**
Instagram, **299, 305**
Islam, **26-29, 68, 119, 418**
Jackson, Andrew, **289**
Jeremiah (Biblical), **30, 31, 233**
Joan of Arc, **82**
Jobs, Steve, **309, 345**
Judaism, **25-29**
Jung, Carl, **196**

Kahneman, Daniel, **242-246, 251**
Kant, Immanuel, **127**
Keynes, John Maynard, **187, 188, 232**
Khaldun, Ibn, **119, 120**
Krugman, Paul, **212**
Kubrick, Stanley, **178, 337, 371**
Lang, Fritz, **343**
Licklider, Joseph, **270**
Locke, John, **95, 124**
Loungani, Prakash, **231**
Lucas, George, **345**
Luddites, **413-416**
Marshall, Alfred, **139, 214**
Marshall, James, **263**
Marx, Karl, **143-145, 161-166, 171, 213, 282, 356, 415**
Max-Neef, Artur Manfred, **310**
McCarthy, Joseph, **281**
McCarthy, John, **371**
McLean, Malcolm, **258, 259**
McLuhan, Marshall, **349**
Méliès, Georges, **342**
Mercantilism, **121, 122, 129**
Mesopotamia, **15, 16**
Meteorological Office, **85**
Milgram, Stanley, **240, 241, 335**
Mincer, Jacob, **165, 166**
Ming (dynasty), **60**
Mischel, Walter, **246**
Moore, Graham, **268**
Moore's Law, **278, 357, 372, 374, 390**
Musk, Elon, **306, 341**
Nabisco, **200**
Nelson, Robert H, **212, 213**
Netflix, **306, 312, 360**
Newton, Isaac, **96, 97, 125, 126, 133, 186, 227**
Nietzsche, Friedrich, **150, 151, 190, 273, 407**

INDEX

Nintendo, 277
Nobel, Alfred, **308**, **429**
Nostradamus, **35**
NW Ayer (company), **201**, **202**
Obama, Barak, 364
Oppenheimer, Robert, **308**
Optimism Bias, **243**, **244**
Oracle (company), 271
Orwell, George, **8**, **332**, **333**, **335**, **380**, **381**, **386**
π (film), **210**
Page, Larry, **284**, **306**
Paine, Thomas, **95**, **127**, **136**
Pavlov, Ivan, **196**
Petty, William, 124
Pinker, Stephen, **387**, **401**
Politics (Aristotle), **117**
Psychohistory (Isaac Asimov), **341**, **379**, **406**
Pythia (Oracle at the Temple of Apollo in Delphi), **19**
RAND Corporation, **176-178**, **270**, **283**
Rational Choice Theory, **164**, **239**, **249**, **283**
Reagan, Ronald, **164**, **182**
Republic (Plato), **117**, **350**
Robber barons, **148**, **180**
Romer, Paul, 212
Samuelson, Paul, **231**
Scott, Ridley, **346**, **426**
Scott, Walter Dill, **197**
Sega, 277
Serra, Richard & Schoolman, Carlota Fay, **304**
Shelly, Mary, **8**, **331**, **380**
Shockley, William, **268**, **269**
Sidewalk Labs, **360**, **361**
Silicon Valley, **137**, **272**, **274-279**, **285-287**, **301**, **305-315**, **324**, **345**, **354**, **377**, **379**, **416**, **429**

Sismondi, Jean Charles Léonard de, **139**
Skinner, BF, **322-324**
Slave trade, **104-107**
Smith, Adam, **127-132**, **136-145**, **156**, **162-166**, **171**, **191**, **242**, **250**, **252**, **260**, **333**, **429**
Snapchat, 299
Socrates, **194**
Sony, 277
Spanish Armada, **67**, **82**
Stalin, Joseph, **159**, **333**
Stanford University, **241**, **243**, **246**, **266**, **270**, **272**, **372**
Stanford, Leland Sr, 266
Sunk Cost Falacy, **211**, **245**, **316**
Sunstein, Cass, 251
System 1/System 2 Thinking, **252**, **307**
Taoism, **26**
Tarkovsky, Andrei, **344**
Tetlock, Philip, **236**
Thaler, Richard, 251
Tian Ming, **40**
Truman, Harry S, **159**
Turing, Alan, **173**, **371**, **372**
Tversky, Amos, **242**
Twitter, **299**, **305**, **321**
Utilitarianism, **140**, **151**, **397**
Verenigde Oost-Indische Compagnie (VOC) (see Dutch East India Company)
Verne, Jules, **8**, **331**, **332**, **343**
Voltaire, **126**, **247**, **424**
Wall Street Crash, **157**, **182**, **231**, **234**, **242**, **284**, **285**
Washington, George, **138**, **182**
Watson, Peter, 233
Watt, James, **100**, **101**
Waverly Labs, **427**
Wells, HG, **8**, **3**, **332**, **280**, **382**

WhatsApp, **299**
Wilde, Oscar, **415, 428**
Wittgenstein, Ludwig, **408, 412**
Wundt, Wilhelm, **195**
Xerox, **271**
YouTube, **300, 321, 393**
Zimbardo, Philip, **241**
Zoroastrianism, **25, 26**
Zuckerberg, Mark, **286, 306**

www.ingramcontent.com/pod-product-compliance
Lightning Source LLC
Chambersburg PA
CBHW070802040426
42333CB00061B/1743